SCIENCE
A CLOSER LOOK

BUILDING SKILLS

Activity Lab Workbook

Macmillan
McGraw-Hill

Contents

Dear Parent or Guardian,

   Today our science class talked about how to work safely when doing laboratory experiments. It is important that you be informed regarding the school's effort to promote a safe environment for students participating in laboratory activities. Please review the safety rules and this entire Safety Contract with your child. This contract must be signed by both you and your child in order for your child to participate in laboratory activities.

Safety Rules:

1. Listen carefully and follow directions.
2. Perform only those experiments approved by your teacher. If you are not sure about something, ask your teacher.
3. Take great care when handling and moving chemicals and hot materials.
4. Conduct yourself in a responsible manner at all times.
5. Always clean up after you have finished an experiment.
6. Always wash your hands before and after an experiment.
7. Do not eat, drink, or chew gum in the laboratory.

Date: _____

   I have read and reviewed the science safety rules with my child. I consent to my child's participation in science laboratory activities in a classroom environment where these rules are enforced.

   Parent/Guardian signature: _____

   I know that it is important to work safely in science class. I understand the rules and will follow them.

   Student signature: _____

Estimados padres o tutor:

Hoy hemos hablado en nuestra clase de Ciencias sobre cómo mantener la seguridad al realizar experimentos científicos. Es importante que ustedes estén informados del propósito de la escuela de promover un entorno seguro para los estudiantes que participan en las prácticas de laboratorio. Por favor, examinen cuidadosamente con su niño o niña las reglas siguientes y el Acuerdo de Seguridad. El acuerdo debe ser firmado tanto por uno de ustedes como por su niño o niña para que él o ella pueda participar en las actividades de laboratorio.

Reglas de Seguridad:

1. Escucha con atención y sigue las indicaciones.
2. Haz sólo los experimentos aprobados por tu maestro o maestra. Pregúntale a él o a ella si no estás seguro de algo.
3. Ejercita sumo cuidado al manipular y transportar productos químicos y materiales calientes.
4. Compórtate en todo momento de manera responsable.
5. No te olvides de limpiar cuando termines de realizar un experimento.
6. Lávate siempre las manos antes y después de hacer un experimento.
7. No comas, bebas ni mastiques chicle en el laboratorio.

Fecha: _____

He leído y examinado las reglas de seguridad de ciencias con mi niño o niña. Doy mi consentimiento para su participación en las actividades del laboratorio de ciencias en un entorno donde se hagan cumplir estas reglas.

Firma de uno de los padres o tutor: _____

Sé la importancia que tiene trabajar con seguridad en la clase de Ciencias. Comprendo las reglas y me comprometo a seguirlas.

Firma del estudiante: _____

What do you know about disease?

Materials
- paper
- pencil

Purpose

Biologists are curious about the natural world and everything that lives in it. Susan Perkins and Liliana Dávolos are biologists at the American Museum of Natural History in New York City. They investigate organisms by looking at their cells under a microscope and analyzing them in the laboratory.

Susan Perkins   Liliana Dávolos

Name _____ Date _____

1 How do people get sick?

_____

_____

_____

_____

_____

2 Do other animals get sick, too?

_____

_____

_____

_____

_____

3 What are some of the same diseases that both people
and other animals get?

_____

_____

Use with **Be a Scientist**

Explore More

4 How do you think scientists study diseases?

_____

_____

_____

_____

Open Inquiry

Think about a time when you got sick. Write a question about how you got sick and why. Think about how you can test your question.

My question is:

_____

_____

How I can test it:

_____

_____

_____

My conclusion is:

_____

_____

_____

_____

Name _____ Date _____

How do diseases spread?

1 How do mosquitoes spread disease?

_____

_____

_____

_____

Materials

• encyclopedia

• Internet

• drawing supplies

2 Besides mosquitoes, what are three other ways that diseases can spread?

_____

_____

_____

_____

_____

_____

_____

3 What are three ways you can help stop the spread of disease?

_____

_____

_____

_____

_____

4 On a separate piece of paper, design and color a poster that shows one way to stop the spread of disease.

© Macmillan/McGraw-Hill

Use with **Be a Scientist**

What are plants and animals made of?

Make a Prediction

Plants and animals are living things. Think about a plant and an animal you have seen. Do you think they are made of similar or different parts?

_____

_____

© Macmillan/McGraw-Hill

Materials

- microscope
- prepared slides of plant leaf cells
- prepared slides of animal blood cells

Test Your Prediction

1 **Observe** Look at the prepared slide of a plant leaf under the microscope. For help using the microscope, ask your teacher.

2 **Record Data** Draw what you see.

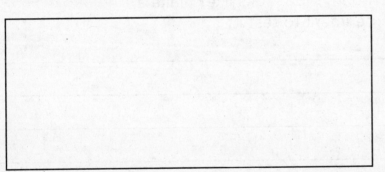

Step 1

3 **Observe** Look at the prepared slide of animal blood under the microscope.

4 **Record Data** Draw what you see.

Step 2

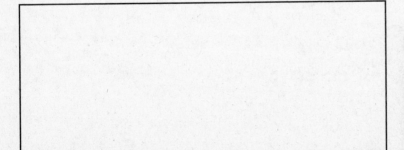

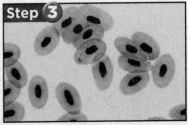

Step 3

Draw Conclusions

5 **Interpret Data** How were the plant leaf slide and animal blood slide alike? How were they different?

_____

_____

_____

6 **Communicate** Write a report explaining whether or not your observations supported your prediction.

_____

_____

Explore More

Examine the drawings you made and think about the living things they came from. Mushrooms are also living things. What do you think a mushroom slide looks like? Make a prediction and plan an experiment to test it.

_____

_____

_____

_____

Open Inquiry

Design an experiment that will compare cells from a number of organisms. Be sure to use living things from each of the six kingdoms: Archaebacteria, Eubacteria, Protists, Fungi, Plants, and Animals.

▶ My question is:

_____

_____

▶ How I can test it:

_____

_____

▶ My results are:

_____

_____

_____

Name _____ Date _____

How are plant cells and animal cells similar?

Materials
- encyclopedias, Internet, or other reference materials
- ruler

Make a Prediction

Plants and animals are made of cells. Do plant cells and animal cells have similar structures?

_____

_____

_____

Test Your Prediction

1 **Research** Look for information about plant cells and animal cells in reference materials. Keep looking until you find written descriptions and diagrams of the cells.

2 On a separate piece of paper, draw a Venn diagram of the parts of animal and plant cells. In the center, write the parts found in both. List the parts and where they are in your diagram here.

_____

_____

3 **Compare and Contrast** What is different about plant cells and animal cells?

_____

_____

Draw Conclusions

4 Was your prediction about the structure of plant cells and animal cells correct? Explain.

_____

_____

Plant and Animal Cells

Materials
- sandwich bags
- sandwich containers with lids
- masking tape (optional)
- pencil (optional)
- gelatin
- vegetables

1 **Make a Model** Put one plastic bag in a storage container. This is a plant cell. Use another plastic bag as an animal cell.

2 Using a spoon, carefully put gelatin in both bags until the bags are almost full.

3 Choose vegetables that look the most like the plant-cell and animal-cell organelles.

4 Place the vegetables that you have picked into the appropriate container and seal the bags.

5 **Compare** Try to stack your models. How well do the plant cells stack compared to the animal cells?

_____

_____

6 **Communicate** Discuss with your classmates which vegetables you selected for your organelles and explain why. Write about one of them.

_____

_____

_____

Experiment

All living things are made up of cells. Every cell has a cell membrane. A cell membrane is a layer around the cell. It lets substances in and out of the cell.

Substances move in or out of a cell depending on their concentrations, or amounts. Substances move from areas where they are crowded to areas where they are less crowded. For example, if a cell has a higher concentration of water than its environment, water will flow out of the cell until the concentration on the inside and outside is balanced. One way to learn more about how cell membranes work is by doing an **experiment**.

▶ **Learn It**

An **experiment** is a test that supports or does not support a hypothesis. To carry out a successful **experiment** you need to perform a test that examines the effects of one variable on another using controlled conditions. You can then use your data to draw a conclusion about whether or not the hypothesis has been supported.

In the following experiment, you will test the effects of various substances on a cell membrane. You will gather and analyze data to support or disprove the following hypothesis: If the concentration of a substance is higher outside the membrane, then the substance will move inside the membrane to balance the concentration.

Name _____ Date _____

▶ Try It

1 Measure two eggs using a balance. Record the measurements in the chart.

2 Pour 200 mL of vinegar into two jars with lids. Carefully lower the two eggs into the jars of vinegar. Tighten the lids and leave the eggs inside for two days.

3 Use a spoon to carefully remove the eggs. Rinse the eggs under water.

4 Measure each egg and record the data in your chart.

5 Pour 200 mL of water into a beaker and 200 mL of corn syrup into another beaker. Carefully lower an egg into each beaker. Leave the eggs inside for one day.

6 Use the spoon to carefully remove the eggs. Rinse the eggs under water.

7 Measure each egg and record the data in your chart.

Materials

- two eggs
- balance
- 2 glass jars with lids
- 400 mL of vinegar
- 200 mL of water
- 200 mL of corn syrup
- spoon
- 2 beakers

	First Measurement	Second Measurement	Third Measurement
Egg #1			
Egg #2			

▶ **Apply It**

8 Now it is time to analyze your data and observations. Use your chart to compare the masses of the eggs.

9 Did the mass of both eggs change? Explain why the masses changed.

_____

_____

_____

_____

10 Does this **experiment** support or disprove the hypothesis?

_____

_____

_____

_____

_____

How can living things be classified?

Purpose

Scientists group, or classify, organisms with certain similarities together. Once an organism is placed into a large group, scientists can then further classify them by the presence or absence of specific characteristics. Compare specimens and classify them based on their characteristics.

Procedure

① **Observe** Look at the specimens your teacher has given you.

② Examine the specimens two at a time and compare them. How are they alike? How are they different? Record your findings in a chart.

Materials

- plant specimens
- rock specimens
- fungal specimens
- animal specimens

Specimens	Similarities	Differences

③ **Classify** Find ways to group the specimens based on their characteristics. For example, you might group the specimens based on whether they move from place to place or whether they take in food or make their own food.

Name _____ Date _____

④ Communicate Compare your classification chart with a classmate's chart. How did your classification methods compare?

_____

_____

Draw Conclusions

⑤ Infer Why do you think classifying organisms helps scientists? Explain.

_____

_____

⑥ Which of the items you classified are more similar, or more closely related, to each other?

_____

_____

_____

Explore More

What other organisms or items can you classify? Observe organisms near your house or your school. Classify them into one of the groups.

_____

_____

_____

_____

Open Inquiry
You can design a classification scheme for the music you listen to, the movies you see, or the television shows you watch. What makes this scheme effective?

▶ My scheme is:

_____

_____

▶ How I can test it:

_____

_____

▶ My results are:

_____

_____

What is living in pond water?

Purpose

In this activity, you will observe living and nonliving things found in pond water.

Procedure

1. Place a drop of pond water on a microscope. Add a cover slip.

2. **Observe** Examine the slide under a microscope. Focus on low power first, and then switch to high power. Move the slide slowly in circles until you see something, and then observe for a minute or two. Keep looking until you have observed several different objects.

3. **Record Data** On a separate piece of paper, draw one example of something that appears to be nonliving. Also draw at least two examples of tiny living organisms that you observed. Label each object as living or nonliving.

_____

_____

_____

Draw Conclusions

4. Explain how you were able to distinguish nonliving objects from living organisms.

_____

_____

© Macmillan/McGraw-Hill

Bread Mold Activity

Materials

- one sandwich bag
- paper
- slice of bread
- graph paper (optional)
- dropper
- water

1 Trace a slice of bread onto graph paper.

2 Put a drop of water on one corner of the bread and put it in a bag. Place the bag in a warm, dark corner.

3 **Observe** When you first see mold, sketch the shape of the moldy area on the graph paper.

4 For three days, use different colors to sketch any new mold growth.

5 **Interpret Data** Each day count the number of squares that are covered with mold.

Day	Number of Squares Covered

6 On a sheet of graph paper create a bar graph to show the growth of the mold from day to day.

Name _____ Date _____

How is water transported in vascular plants?

Form a Hypothesis

All vascular plants have vessels that transport food and water in the plant. How does the amount of leaves on a plant affect transport through a plant stem? Write your answer as a hypothesis in the form "If the number of the leaves on a plant decreases, then . . ."

_____

_____

Materials

- 3 plastic cups
- water
- blue food coloring
- 3 celery stalks with leaves
- ruler

Test Your Hypothesis

1 Fill 3 plastic cups with water. Be sure that each cup has the same amount. Put 3 drops of food coloring in each cup of water.

2 Break all the leaves off one celery stalk. Remove all but one leaf on another stalk. Leave the third stalk intact. Place a celery stalk in each cup.

Step **1**

3 **Observe** On the following day, examine each cup. What happened to the water? Note any changes.

_____

_____

Step **2**

4 **Measure** Use a ruler to measure how far up the water traveled in each celery stalk.

_____

_____

Draw Conclusions

5 What are the independent and dependent variables in this experiment?

_____

_____

6 **Interpret Data** Did the amount of leaves affect the transport of water?

_____

_____

7 Did your results support your hypothesis?

_____

_____

Explore More

What other variables can affect the movement of water through a plant? How will adding sugar or salt affect water transport in a plant? Form a hypothesis and test it. Then analyze and write a report of your results.

_____

_____

_____

Name _____ Date _____

Open Inquiry

Design an experiment to add color to the petals of a white carnation. Formulate a hypothesis and then design and carry out an experiment to test it.

▶ My question is:

_____

_____

▶ My hypothesis is:

_____

_____

_____

▶ My results are:

_____

_____

What is inside the strings of a celery stalk?

Materials
• Internet, encyclopedias, or other reference materials • Books

Form a Hypothesis

A celery stalk contains many long strings. The strings are thickest at the wide end of the stalk. Write a hypothesis about what these strings do.

_____

_____

Test Your Hypothesis

1. **Research** Search on the Internet for information about celery strings. Also read about transport in plants in other science reference books.

2. **Summarize** What did you find out about celery strings?

_____

_____

_____

_____

3. **Draw** On a separate piece of paper, make a drawing of what is inside a celery string.

_____

_____

_____

Draw Conclusions

4. What are the two functions of the strings in celery stalks?

_____

_____

Name _____ Date _____

Observe a Root

Make a Prediction

Materials
- carrot
- knife

1 **Observe** Look at a carrot cut lengthwise. What structures do you see?

_____

_____

2 Look at a cross section of a carrot. Can you identify the epidermis, cortex, and inner transport layers?

3 Draw a diagram of the carrot in cross section.

_____

| |
| |

4 **Infer** Is the carrot a fibrous root or a taproot?

_____

5 Would it be easier to pull a plant with a taproot from the ground or a plant with a fibrous root system? Explain your answer.

_____

_____

How do you classify animals?

Purpose

A dichotomous key lists traits of organisms. It gives directions that lead you to an organism's identity. Create a dichotomous key to identify the animals shown.

Procedure

1 **Observe** Look at the animals shown. What features do they have? How can you use these features to classify them?

_____

_____

2 Make a key. The key should include a series of yes or no questions that can help you identify the animals.

Question	Yes	No
Is this animal able to move by itself from place to place?		

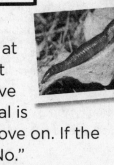

3 Start your key with a general question. Make each additional question eliminate at least one animal. For example, you might start by asking: Is this animal able to move by itself from place to place? If the animal is a worm, you would answer "Yes," and move on. If the animal is a sponge, you would answer "No."

④ Keep writing questions until you can single out one animal in your key.

⑤ **Communicate** Exchange keys with a partner. Use their key to identify an animal.

Draw Conclusions

⑥ **Infer** Could you use your key to identify other animals? Explain.

_____

_____

Explore More

How would you change your key to make it more useful? Which questions would you change? Which questions would you keep the same?

_____

_____

_____

_____

Open Inquiry

Think about different ways you could classify animals. You may want to design your own kingdoms.

▶ My question is:

_____

_____

▶ How I can test it:

_____

_____

▶ My results are:

_____

_____

_____

How can you identify vertebrates with a field guide?

Materials
- field guides from the library

Purpose

When you see an animal, how do you know it is a vertebrate? How do you know what kind of vertebrate it is? In this activity, you learn how to use a field guide to identify vertebrates.

Procedure

1 **Observe** Look carefully at an animal's shape and body structures. Study the animals in your field guide.

2 **Compare and Contrast** What do the animals all have in common? What differences do you see among them?

_____

_____

_____

3 **Infer** Are the animals shown here vertebrates or invertebrates? Explain.

_____

_____

4 **Identify** Observe another vertebrate, in a photo or on a nature walk. Determine what kind of vertebrate the animal is. Then look in a field guide to identify the animal. Study the descriptions and photos given in the field guide section on the kind of vertebrate you observed.

Draw Conclusions

5 Decide which animal shown is the same as yours. What are its common and scientific names?

_____

Draw Conclusions

6 Which parts of your model represent bones?
Which part represents muscle?

_____

_____

7 **Infer** Which muscle in your body is similar to this model?
Explain.

_____

_____

8 How do muscles work? What happens when muscles
shorten and lengthen? Explain.

_____

_____

Explore More

What would happen if you did not cut a slit in the straw?
Make a prediction and plan an experiment to test it.

_____

_____

_____

_____

Name _____ Date _____

Open Inquiry

Think about how you could model the connection between two bones. Write a question about how bones connect and how you would model it.

▶ My question is:

_____

_____

▶ How I can test it:

_____

_____

▶ My results are:

_____

_____

Can one muscle move a bone up and back?

Materials

- encyclopedias, Internet, or other reference materials
- string
- scissors

Make a Prediction

Muscles cause bones to move. Think about what happens when you bend your arm at the elbow and then straighten it. Did the same muscle cause both movements? Make a prediction.

_____

_____

Test Your Prediction

1. **Research** Look for information about how muscles produce the movements of the body. Focus on how the forearm is moved by bending at the elbow.

2. **Use a Model** Work with a partner to model how muscles work. One partner will provide the force to move the other partner's arm by causing the elbow to flex.

3. Tie string around your partner's wrist. Pull on the string to model the force created by a muscle. Use one piece of string for each direction of movement.

Draw Conclusions

4. Can one muscle move the arm in opposite directions? Explain.

_____

_____

_____

_____

Name _____ Date _____

Tiny Filters

Materials

- water
- paper towel
- pepper
- sugar

1 Mix pepper and water in a cup.

2 Pour the mixture through a paper towel into another cup. What went through the paper towel and what did not?

_____

_____

3 Mix sugar and water in a cup.

4 **Experiment** Pour the mixture through a paper towel into another cup. What went through the paper towel and what did not?

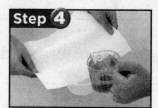

_____

_____

5 How would you explain what you observed?

_____

_____

_____

6 **Infer** How is the paper towel similar to nephrons?

_____

_____

_____

_____

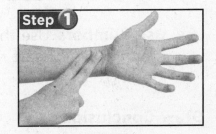

Be a Scientist

Structured Inquiry

When does your heart work the hardest?

Materials
- stop watch
- graph paper

Form a Hypothesis

Driven by the heart, your blood cells travel around your body carrying oxygen to your body cells. When you exercise, your body requires more oxygen. What happens to your heart when you exercise? Write your answer as a hypothesis in the form "If the body requires more oxygen, then . . ."

_____

_____

Test Your Hypothesis

1. **Experiment** Take your pulse when you are resting. Press lightly on the skin of your wrist until you feel a beat. Then count how many beats you feel in 30 seconds. Record your pulse in a chart.

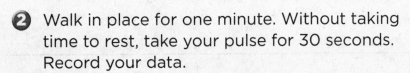

Step 1

2. Walk in place for one minute. Without taking time to rest, take your pulse for 30 seconds. Record your data.

Step 2

© Macmillan/McGraw-Hill

3 Jog in place for one minute. Without taking time to rest, take your pulse for 30 seconds. Record your data.

Activity	Pulse One	Pulse Two	Pulse Three	Pulse Four	Pulse Five
Resting					
Walking					
Jogging					

4 Rest for 5 minutes then repeat steps 1–3 four more times.

5 **Use Numbers** Use the data you collected to make a graph of your heart rate when resting, walking, and jogging.

Draw Conclusions

6 **Interpret Data** Did your pulse change as your activity level increased?

_____

7 Did the experiment support your hypothesis? Explain.

_____

_____

_____

8 What were the independent and dependent variables?

_____

_____

Guided Inquiry

When do your lungs work the hardest?

Form a Hypothesis

You have already tested the effects of exercise on your heart.
Do you think exercise affects your lungs? What happens to
your breathing rate when you exercise? Write your answer as
a hypothesis in the form "If the body requires more oxygen,
then . . ."

_____

_____

Test Your Hypothesis

Design a plan to test your hypothesis. Then write out the
materials, resources, and steps you need. Record your results
and observations as you follow your plan and conduct your
experiment.

_____

_____

_____

_____

Draw Conclusions

Did your experiment support your hypothesis? Why or why
not? Present your results to your classmates.

_____

_____

_____

_____

Open Inquiry

How are your heart rate and your breathing rate affected by other activities? Would your heart rate or breathing increase if you read a book, play a sport, or try to solve a math problem? Determine the steps you would follow to answer your question. Record and document the resources you would use during your investigation.

▶ My question is:

_____

_____

_____

▶ How can I test it?

_____

_____

_____

▶ My results are:

_____

_____

_____

© Macmillan/McGraw-Hill

Can some flowering plants grow without seeds?

Make a Prediction

You have learned that flowering plants use seeds to reproduce. Can some flowering plants reproduce without seeds? Can you use part of a plant to create a new plant? Make a prediction.

_____

_____

_____

_____

Materials

- philodendron plant
- safety scissors
- hand lens
- plastic cup
- water
- 2-week-old cutting in a plastic cup (optional)

Test Your Prediction

1. Cut a piece of stem from the philodendron plant that measures 15 cm (6 in.) in length. Cut off the leaves that are closest to the plant. Leave 2 leaves at the very tip of the cutting.

Step **1**

Step 3

2 Observe Look at your cutting with the hand lens. Record your observations.

_____

_____

3 Fill the plastic cup 2/3 of the way with water. Place the cutting into the plastic cup

4 Interpret Data Examine your cutting each day with the hand lens. Record your observations and any changes.

_____

_____

Draw Conclusions

5 **Infer** What happened to the cutting in the plastic cup with water?

_____

6 Is it possible to grow a new plant without planting a seed? Explain.

_____

_____

Explore More

Could other plants grow in a way that is similar to the philodendron plant? Plan an investigation to answer the question. Write a report of your results and present them to the class.

_____

_____

_____

Open Inquiry

Design an experiment to see if plants can grow without seeds.

My question is:

_____

How I can test it:

_____

_____

My results are:

_____

Name _____ Date _____

Can a new plant grow from a potato?

Materials

- potato with eye
- hand lens
- 4 toothpicks
- clear plastic cup

Make a Prediction

Write a prediction about if and how a new plant can grow from a potato.

_____

_____

Test Your Prediction

1. **Observe** Observe the potato with a hand lens. Locate any small indentations called eyes.

2. Place four toothpicks around the center of the potato. Do not poke a toothpick into any of the eyes.

3. Half fill the cup with water. Place the potato in the cup so that the toothpicks hold the potato about half way in the water.

4. **Communicate** Describe the potato as it looks now.

_____

_____

5. **Observe.** Observe your potato for about 5 minutes each day for about two weeks. Make sure to keep water in the cup. Record your observations every day.

Draw Conclusions

6. Was your prediction correct? Did a new plant grow from the potato?

_____

_____

Asexual Reproduction Poster

1 Research three types of asexual reproduction. Use the Internet, books, and magazines as sources.

2 Find out what organisms use these types of asexual reproduction.

3 Make a poster that compares the types of asexual reproduction you researched. Your poster can be a chart, graph, diagram, or table.

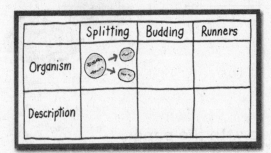

4 **Communicate** Make illustrations or cut out pictures of organisms that use these types of asexual reproduction. Place them in your poster and describe them.

5 How are the types of asexual reproduction similar and different?

_____

_____

_____

_____

Name _____ Date _____

How do flowering plants reproduce?

Form a Hypothesis

Do seeds require moist or dry conditions to grow? Write your answer as a hypothesis in the form, "If seeds are placed in a moist environment, then . . ."

_____

_____

_____

Materials

• paper towels

• 4 plastic cups

• water

• lima bean seeds

• potting soil

Test Your Hypothesis

1 Moisten a paper towel and place it at the bottom of a plastic cup.

2 Place 1 lima bean seed inside the cup with the paper towel. Fold the paper towel over the seed.

3 Repeat steps 1 and 2 with a dry paper towel.

Step **2**

© Macmillan/McGraw-Hill

4 **Experiment** Place the cups in a sunny spot and observe them daily for 5 days. Record your observations in a data table.

	Moist	Dry
Monday		
Tuesday		
Wednesday		
Thursday		
Friday		

5 After 5 days, fill 2 cups with potting soil.

6 Take each seed and place it into a cup with soil. Gently cover each seed with soil and sprinkle some water on top.

Step **6**

7 **Observe** Place the cups in a sunny spot. Water the seeds daily and look for any changes. Record your observations in your data table.

	Moist	Dry
Monday		
Tuesday		
Wednesday		
Thursday		
Friday		

© Macmillan/McGraw-Hill

Name _____ Date _____

Draw Conclusions

8 What were the independent and dependent variables in this experiment?

_____

_____

9 **Infer** What conditions were needed for your seeds to grow?

_____

10 Did your results support your hypothesis?

_____

Explore More

Keep observing your plant over time. What does your plant need to produce seeds? Make a prediction. Test your prediction and present your results to the class.

_____

_____

Open Inquiry

Design an experiment to see if seeds are important to a type of plant.

My question is:

_____

_____

How I can test it:

_____

_____

My results are:

_____

Which seed sprouts first?

Form a Hypothesis

Which seed do you think will sprout first? Answer the question with a hypothesis in this form, "If all the seeds are given the same amount of water and sunlight, then . . ."

_____

_____

_____

_____

Materials
• bean seed
• corn seed
• sunflower seed
• 3 paper towels
• 3 plastic cups

Test Your Hypothesis

1. Place a crumpled piece of paper towel in each cup. Add water to each cup to moisten the paper towel.

2. Place a sunflower seed in one cup, a bean seed in another cup, and a corn seed in the third cup. Moisten a piece of paper towel and cover each seed with the moist towel.

3. Set the seeds in a sunny location. Check your seeds each day to find out which seed sprouts first. Make sure to keep the paper towel in each cup moist.

4. **Communicate** Record your daily observations in a chart.

Draw Conclusions

5. Which seed sprouted first? Was your hypothesis correct?

_____

_____

Name _____ Date _____

Comparing Seeds

Materials
- seeds

1 **Observe** Take a look at each seed type.

2 Record the characteristics of each seed in a table. Use the following headings: size, shape, weight, toughness

	seed 1	seed 2	seed 3	seed 4
size				
shape				
weight				
toughness				

3 **Predict** How do you think each of the seeds you observed gets dispersed? Explain your answer.

_____

_____

Observe

You just learned about plant life cycles and plant structures. For example, flowering plants reproduce sexually by forming seeds when sperm from pollen fertilizes an egg cell inside the pistil.

Perfect flowers have both a pistil (female part) and stamen (male part). Imperfect flowers have either a pistil or stamen, but not both. How do scientists know this? They **observe** real flowers!

▶ **Learn It**

When you **observe**, you use one or more of your senses to learn about an object. It is important to record what you **observe**. One way is to draw a diagram with labels to identify exactly what you saw. You can record other observations, such as odors and sounds, under the diagram. Then you can use the information to help identify other plants and their parts.

This diagram is a record of someone's observations. Each flower part is labeled. Note the observations under the diagram.

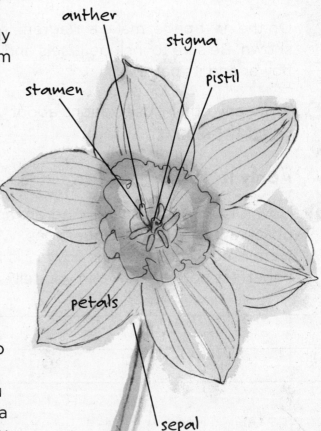

anther

stigma

pistil

stamen

petals

sepal

Petals feel smooth and velvety to the touch. Pollen feels like soft powder. Petals have no smell. Plant parts make a crisp "SNAP" when broken.

Name _____ Date _____

▶ **Try It**

1 **Observe** Look at a flower.

2 On the next page, make a diagram like the one shown. Be sure to include labels and color your flower and its parts.

3 Write any other observations about your flower under the diagram.

▶ **Apply It**

1 Now it is time to use your diagram and other observations to answer questions. Which senses were used to observe this flower? Is this a perfect or imperfect flower? How can you tell?

_____

_____

_____

Materials

• flower

• paper

• pencil, colored pencils or markers

• hand lens

© Macmillan/McGraw-Hill

_____
_____
_____

2 Continue to use your observation skills. Choose an object in your classroom, such as a stapler, pencil sharpener, TV, or the intercom system.

3 **Observe** Look at the object you chose. Then make a diagram of the object on another sheet of paper. Include labels to identify any parts and how they are used. Write any other observations, such as how it feels to the touch or the sound it makes, under the diagram.

4 Share your diagram and observations with your classmates.

Name _____ Date _____

What are the stages in an animal's life cycle?

Purpose

You are part of an expedition that is studying the life cycle of frogs. You have collected some data about the frogs you observed. Interpret your data and photos to determine how long each stage of a frog's life cycle lasts.

Stage 1: Fertilized eggs
Date: April 1

Stage 2: Tadpole
Date: April 5

Procedure

1. **Observe** Take a look at the stages involved in frog development.

2. On the next page, create a table to record changes in the frog's body structure during each stage of development.

Stage 3: Tadpole
Date: June 23

Stage 5: Mature Frog
Date: July 21

Stage 4: Froglet
Date: July 7

© Macmillan/McGraw-Hill

Stages in a Frog's Life				
	Fertilized eggs	**Tadpole**	**Froglet**	**Mature frog**
Length of stage				
What it looks like				

③ **Interpret Data** Use the photos on page 50 to determine how long each stage lasts. Record the information in your chart.

Draw Conclusions

④ What was the shortest stage in frog development? What was the longest stage?

_____

_____

⑤ **Infer** When did the organism seem to change the most?

_____

Name _____ Date _____

Draw Conclusions

6 How is the organism in Stage 2 different than the organism in Stage 4?

_____

_____

_____

Explore More

How does the fertilized frog egg develop into a tadpole? Use the Internet or other sources to find photographs of the first four days of a tadpole's life. Describe the changes you see.

_____

_____

_____

Open Inquiry

Design an experiment to find out at which stage a tadpole turns into a frog.

My question is:

_____

_____

How I can test it:

_____

_____

My results are:

_____

_____

How does a grasshopper develop?

Materials

- encyclopedia
- reference books
- magazines

Purpose

Research how a grasshopper develops and changes during its lifespan.

Procedure

1. Use encyclopedias, other reference books, or the Internet to find out how grasshoppers change from when they hatch from an egg until they are adults.

2. **Communicate** Make drawings in the space below that show how the grasshopper changed. Label your drawings.

Model External Fertilization

Materials

• fish tank

• fish tank gravel

• 15 blue marbles

• 15 red marbles

1. **Make a Model** Cover the bottom of a tank with about 1 cm (1/2 in.) of gravel. Fill the tank about 2/3 full with water.

2. Scatter 15 blue marbles in the water. The blue marbles represent unfertilized eggs.

3. After the blue marbles settle, scatter 15 red marbles in the tank.

4. How many blue marbles got "fertilized", or touched, by red marbles?

   _____

   _____

5. **Infer** What does this tell you about the accuracy of external fertilization?

   _____

   _____

Structured Inquiry

How does light affect the life cycle of the wax moth?

Form a Hypothesis

Some insects have different stages in their life cycle in which the body looks nothing like the adult stage. This is called complete metamorphosis. For example, a butterfly's larval stage is a caterpillar. A caterpillar does not have wings and looks very little like the adult. Wax moths have a similar life cycle.

How does the amount of light affect the wax moth's life cycle? Write your answer as a hypothesis in the form "If the amount of light is decreased, then wax-worm larvae will . . ."

_____

_____

Materials

- 2 plastic cups with lids
- wax-worm food
- forceps
- wax worms
- black construction paper
- lamp
- petri dish
- metric ruler

Test Your Hypothesis

1 Create wax-worm habitats by filling two plastic cups half way with wax-worm food. Use forceps to place 5 wax worms in each cup. Then put the lids on each cup.

© Macmillan/McGraw-Hill

2. Tape black construction paper around one of the cups. Be sure to cover the cup completely.

Step 2

3. **Use Variables** Place the uncovered cup under a lamp or a bright window. Place the covered cup away from the light source.

Step 3

4. **Measure** Remove a wax worm from one of the cups and place it on a petri dish. Use a metric ruler to measure the length and width of the wax worm. Be sure to place the larvae back in the correct cup.

5. Repeat step 4 with each of the wax worms. Record the measurements in a data table.

Wax Worm Growth					
Original size					
Amount of growth					
Amount of growth					
Amount of growth					

6 **Observe** Measure the wax-moth larvae every two days until they pupate. Record the measurements in your data table. Record any other changes you see.

Draw Conclusions

7 Did the growth data of wax-moth larvae under the lamp source differ from the larvae in the covered cup?

_____

_____

8 **Infer** Did the amount of light in the wax-worms' environment have any effect on their life cycle?

_____

_____

_____

Guided Inquiry

How does temperature affect the life cycle of the wax moth?

Form a Hypothesis

Does temperature affect the rate of the wax-worm's life cycle? Write your answer as a hypothesis in the form "If the temperature of the environment of wax-moth larvae is increased, then wax-moth larvae will . . ."

_____

_____

Name _____ Date _____

Test Your Hypothesis

Design an experiment to test your hypothesis. Write out the materials you need and the steps you will follow. Record your results and observations.

Draw Conclusions

Did your results support your hypothesis? Why or why not? Present your results to your classmates.

Open Inquiry

What else can you learn about insect life cycles? For example, how does light or temperature affect the life cycle of a butterfly?

Design an experiment to answer your question. Your experiment must be organized to test only one variable. It must be written so that other students can complete the experiment by following your instructions.

My hypothesis is:

_____

_____

How can I test it?

_____

_____

My conclusions are:

_____

_____

What are some inherited human traits?

Purpose
Everyone has different physical characteristics, but some traits are similar in different people. Do you have any of the same traits as your classmates? Perform a survey to find out which traits your classmates possess. Use your data to determine which traits appear most frequently.

Materials
• paper
• pencils

regular thumb

hitchhiker's thumb

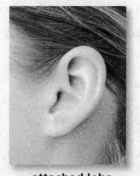
attached lobe

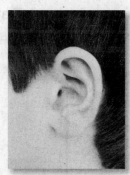
unattached lobe

non-rolled tongue

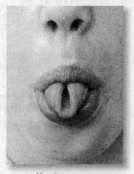
rolled tongue

© Macmillan/McGraw-Hill

Name _____ Date _____

Procedure

1 Have a partner check you for each of the traits shown. Record which form of the trait you have in the chart.

	hitchhiker's thumb	regular thumb	attached earlobe	unattached earlobe	rolled tongue	nonrolled tongue
Yes						
No						

2 Reverse roles and repeat.

3 **Communicate** Tally your results and combine them with your classmates' results in a classroom chart.

4 **Interpret Data** Use the data from the classroom chart to create a bar graph of your results.

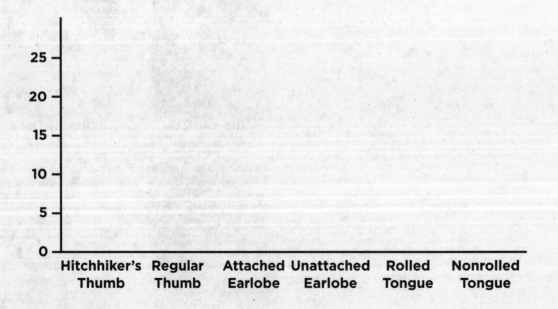

Draw Conclusions

5 **Use Numbers** Find the percent of each trait in the class.

6 Which form of each trait appears more frequently?

_____

_____

7 **Infer** Are some forms of traits more common than others? If so, why?

_____

_____

_____

Explore More

How do your results compare with a larger group? Plan an experiment that would answer this question.

Open Inquiry

Design an experiment to find out about other traits.

My question is:

_____

How I can test it:

_____

My results are:

_____

_____

Name _____ Date _____

Which trait occurs more often?

Materials

• coffee can

• 15 red marbles

• 5 green marbles

Make a Prediction

Assume that red marbles represent a dominant trait and green marbles represent a recessive trait. Which trait do you think occurs more often? Write a prediction.

_____

_____

Test Your Prediction

1 Put all the marbles in the coffee can and stir them around. Without looking, take five marbles out of the can. How many of each color did you select?

_____

2 Continue taking marbles out of the coffee can until you have removed 10 marbles. How many of each color did you select?

_____

3 Repeat the process one more time until you have taken 15 marbles out of the can. How many of each color did you select?

_____

4 Communicate Based on the number of each color marble you selected, what did you learn about dominant and recessive traits?

_____

5 Communicate Was your prediction correct?

_____

Inherited Traits in Corn

Each corn kernel is a separate seed that inherits traits, such as kernel color, from a parent plant.

1 **Observe** Look at an ear of corn. What do you notice?

_____

2 Count the number of purple kernels on your ear of corn. Record the number.

_____

3 Count the number of yellow kernels on your ear of corn. Record the number.

_____

4 **Interpret Data** Which color kernel occurs more frequently?

_____

5 Is the trait for purple kernels dominant or recessive? Explain.

_____

_____

Name _____ Date _____

How do organisms in a food chain interact?

Purpose

A food chain models how food energy is transferred from one organism to another. Producers make their own food. Herbivores consume producers. Carnivores consume herbivores. Create food chains using the list below.

Materials

- blank note cards
- construction paper
- markers
- magazines
- scissors
- glue stick

PRODUCERS	HERBIVORES	CARNIVORES
algae	grasshopper	wolf
berries, flowering plant	deer	otter
shrub	chipmunk	hawk
seeds, grass	squirrel	robin
acorn, oak tree	fish	owl

Procedure

1. Make cards for the organisms listed in the chart above. Draw or glue a picture of an organism on each card.

2. Create a three-column chart on the paper. Label the columns as shown.

Step 2

First Level | Second Level | Third Level
1. Acorn | Squirrel | Owl
2.

3. Use your organism cards to make five food chains. Place the organism cards on your chart under the correct columns.

Draw Conclusions

4 **Communicate** Compare your food chains with a classmate's food chains. Explain how they compared.

_____

_____

_____

_____

_____

5 **Infer** Can food chains overlap? How does this affect the ecosystem?

_____

_____

_____

_____

_____

Explore More

Research one of your food chains. What ecosystem is it part of? What other organisms are part of this ecosystem? How are these organisms connected to your food chain?

_____

_____

_____

_____

Open Inquiry

How do food chains differ in other ecosystems? Think about different kinds of ecosystems and design an experiment to compare the ecosystem you just studied with this new ecosystem. Are they really different?

My question is:

_____

How I can test it:

_____

_____

_____

My results are:

_____

_____

_____

What are some food chains in a desert?

Purpose
Illustrate a food chain from a desert ecosystem.

Procedure

1 Use encyclopedias, reference books, or the Internet to research what organisms live in a desert and what food chains they form.

2 **Communicate** What are the names of some plants and animals that live in the desert?

_____

_____

_____

3 **Communicate** Draw a simple food chain from the desert ecosystem you researched.

Materials

• encyclopedia

• science reference books

• Internet

• colored pencils

Draw Conclusions

4 What organisms are producers in a desert food chain? What organisms are consumers?

_____

_____

Name _____ Date _____

Energy Transfer

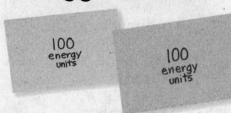

Materials

• Index cards

• markers

• poster paper

❶ Make 100 energy cards. Each card represents 100 energy units.

❷ Make an energy-level poster with four levels: Producers, Herbivores, Carnivores, and Top Carnivore.

❸ Place 100 cards on the Producers. How many total energy units does this level have?

_____

❹ Continue to take 10% of the energy units from one level to the next level up to the Top Carnivore level. Use scissors if necessary.

❺ **Draw Conclusions** How many energy units does the Top Carnivore level have?

_____

© Macmillan/McGraw-Hill

What do organisms need to survive?

Make a Prediction

What do organisms need to survive? Do organisms in an aquatic environment need different things than organisms in land environments? Make a prediction.

_____

_____

_____

Test Your Prediction

1 Make a water environment. Place gravel in one container. Fill the container with pond water. Add water plants and snails.

2 Make a land environment. Place gravel in the other container and cover it with a layer of soil. Add grass seeds and earthworms and cover them with additional soil. Water the seeds.

Materials

- gravel
- 2 containers with lids
- pond water
- water plants
- water snails
- soil
- grass seed
- earthworms

Step **2**

© Macmillan/McGraw-Hill

Name _____ Date _____

3 Cover each container with a lid. Place the containers in a well-lit place out of direct sunlight.

Step **3**

4 **Observe** Examine your containers for changes every day for a week. Do the organisms in each environment interact? Record your observations.

_____

_____

_____

_____

Draw Conclusions

5 What are the abiotic and biotic parts of water and land environments?

_____

_____

6 **Infer** How do the plants help the animals survive in the water environment? The land environment?

_____

_____

7 What would happen to each environment if the plants or animals were removed?

_____

_____

_____

Explore More

What other factors affect an organism's survival? Try adding more plants or animals to your environments. Try placing your environments in the dark for a few days. How do the environments change?

_____

_____

_____

_____

_____

Open Inquiry

Animals and plants need many factors from their ecosystems to survive. Design an experiment in which you determine whether or not an animal needs something from the environment to survive.

My question is:

_____

How I can test it:

_____

_____

_____

My results are:

_____

_____

_____

What changes happen in environments with biotic factors?

Make a Prediction

Make a prediction about how an environment will change with biotic factors added to it.

_____

_____

Test Your Prediction

1 Add a thin layer of gravel to the bottom of each cup. Add pond water to each cup.

2 △ **Be Careful.** Add a pond snail and a piece of elodea plant to one cup. Cover both cups with plastic wrap. Place both cups in an area with no direct sunlight.

3 **Communicate** Observe your ecosystems each day for at least five days. Record any changes that happen in the environments on a chart.

Draw Conclusions

4 **Compare** What changes occurred in both environments? In only one environment?

_____

_____

5 Was your prediction correct? Explain.

_____

_____

Limiting Factors

Materials

- construction paper
- scissors
- metric ruler

1 ⚠ **Be Careful.** Use scissors to cut out 20 2.5 cm (1 in.) circles. Each circle represents the range that the roots of the plant extend.

2 **Measure** Create an environment for these plants by making a 20 cm (8 in.) square box on your desk.

3 Toss 8 plants into the environment. If a plant does not touch another plant, it "survives." If the plant touches another plant, remove the plant and any plant that it touches. Record your results in a data table.

Number of Plants in the Environment	Number of Plants that Survived

4 Increase the number of plants that you toss: 10, 12, 14, and so on. Record your results. Which number of plants tossed allows the most plants to survive?

_____

_____

5 **Infer** How can crowding be a limiting factor for a population?

_____

_____

Name _____ Date _____

Predict

You just read about how some organisms get food by eating other organisms. Can anyone know in advance what effect this will have on the population size? When scientists have questions like that, they conduct simulations and study the results. Then they can **predict** what might happen in a similar situation.

Materials

- masking tape

- eight 7.5 cm cardboard square

- 100 2.5 cm construction paper squares

- graph paper

▶ **Learn It**

When you **predict**, you state the possible results of an event or experiment. Then you conduct a test and interpret the results to determine if your prediction was correct.

It is important to record your predictions, as well as any measurements or observations you make during the test. Your observations and measurements provide written proof of whether or not your prediction was correct. In this activity, you will predict how population size will change.

▶ **Try It**

How many deer do you **predict** will survive in a population of wolves? Use what you have learned about predators and prey to write your prediction. Then use the model to see if your prediction was correct.

① Use tape to mark off a 60 cm by 60 cm square. This square represents a meadow. Distribute 10 of the 2.5 cm paper deer squares in the meadow.

② Toss the 7.5 cm cardboard wolf square in the meadow. Remove any deer that touch the wolf. In order to survive, the wolf must catch, or touch, 3 deer.

Step 2

If the wolf survives, it produces 1 offspring. If the wolf does not catch any deer, it starves.

3 Record your results in the data table. What happened to the wolf and deer in this trial?

_____

_____

_____

Predator-Prey Results							
Trial	Deer	Wolf	Deer Caught	Wolves Starved	Wolves Surviving	New Baby Wolves	Deer Left
1							
2							
3							
4							
5							
6							
7							
8							
9							
10							
11							
12							
13							
14							

④ At the start of the next trial, double the deer remaining from the first trial to represent new deer offspring. Disperse these new deer in the meadow.

⑤ If the entire deer population was caught by the wolf in the previous trial, then add 3 new deer to the meadow.

⑥ In each additional trial throw a wolf square once for each wolf. This includes any surviving wolves from previous trials and any of the offspring produced in previous trials.

⑦ Record your results in your data table. Repeat steps 1 through 6 for a total of 14 trials.

▶ **Apply It**

Predict the outcomes for 6 trials. Base your prediction on the pattern you observed during the first 14 trials. Then actually model trials 15 to 20. Were your predictions correct?

_____

_____

_____

_____

Graph the data for your 20 trials on a sheet of graph paper. Place the deer and wolf data on the same graph so that the interrelationship can be easily observed. Label the vertical axis "Number of animals" and the horizontal axis "Trials." Use one color for the deer data and another for wolf data.

How do adaptations help animals survive in their environment?

Materials

- sow bugs
- tray
- hand lens
- paper towels
- water

Form a Hypothesis

Sow bugs are animals that live under logs, leaves, and rocks. Are sow bugs adapted to prefer damp or dry environments? Write your answer as a hypothesis in the form "If moisture in the sow bug's environment is increased, then . . ."

_____

_____

_____

Test Your Hypothesis

1. **Observe** Place 15 sow bugs on the tray. Examine the sow bugs with the hand lens. Record your observations.

Step 1

_____

_____

_____

Name _____ Date _____

2 **Experiment** Tear four paper towels in half. Make sure they are the same size. Dampen two of the halves.

3 Move the sow bugs to the center of the tray. Place the moist paper towels in one end of the tray. Place the dry paper towels on the opposite side of the tray.

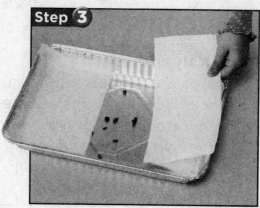

Step 3

4 Watch the sow bugs for several minutes. Look for changes in their behavior.

5 After 10 minutes, count the sow bugs on each side of the tray. Record your results.

⚠ **Be Careful.** Wash your hands after handling sow bugs.

_____

Draw Conclusions

6 Based on your observations, what traits help sow bugs survive in their environments?

_____

7 What were the independent variable and dependent variable? What variables remained constant?

_____

_____

_____

_____

8 **Infer** Did your results support your hypothesis? Explain why or why not.

_____

Explore More

Are sow bugs adapted to prefer dark or light environments? Form a hypothesis and test it. Then analyze and write a report of your results.

_____

_____

_____

Open Inquiry

We've seen that sow bugs prefer moist environments to dryer ones. Are there other environmental factors they prefer? Design an experiment and come up with a plan to test your hypothesis.

My hypothesis is:

_____

How I can test it?

_____

_____

_____

My results are:

_____

_____

_____

Name _____ Date _____

How is an earthworm adapted to its environment?

Make a Hypothesis

Earthworms live underground, where it is dark and moisture keeps their skin moist. How do you think an earthworm responds to light? Write a hypothesis.

_____

_____

Test Your Hypothesis

1 Observe Place moist paper towels on the bottom of a flat plastic container. Place an earthworm in the center of the paper towels. What is the worm doing? How does it move?

_____

_____

2 Place a piece of black paper over half of the container. Observe how the earthworm responds to this change, and record your observations.

_____

_____

Draw Conclusions

3 Interpret Data Did your experiment support your hypothesis about how an earthworm responds to the environment? Explain.

_____

_____

_____

Name _____ Date _____

Leaf Adaptations

Materials
- oak leaf
- *Elodea* leaf
- pine needles
- colored pencils
- ruler

1 Examine an oak leaf, pine needles, and an *Elodea* leaf. Draw what you see.

2 **Measure** Use a ruler to measure each leaf you draw. Record your data.

_____

3 Break open each leaf. How do the leaves compare?

_____

_____

_____

4 **Infer** Which environments are each of the leaves adapted to? Explain.

_____

_____

_____

_____

_____

© Macmillan/McGraw-Hill

Name _____ Date _____

How do water droplets form?

Form a Hypothesis

Water droplets occur when water changes from a gas to a liquid. Does temperature affect water droplet formation on an object? Write your answer as a hypothesis in the form "If the temperature of a glass is decreased then . . ."

_____

_____

_____

Materials

- 2 glasses
- ice
- food coloring
- water
- spoon
- salt
- 2 saucers

Test Your Hypothesis

1. Fill one glass completely with ice. In a separate glass, add a few drops of food coloring to some cold water and stir. Then pour the water into the glass that is full of ice.

2. Fill an empty glass with room-temperature water. Add a few drops of food coloring to the water and stir. Be sure to use the same amount of food coloring and water in each glass.

3. **Experiment** Sprinkle salt onto each saucer. Then put one glass on each saucer. Allow the glasses to sit for 30 minutes.

4. **Observe** What do you see on the sides of each glass?

_____

_____

Use with **Lesson 1**
Cycles in Ecosystems

© Macmillan/McGraw-Hill

Draw Conclusions

5 What does the color of the droplets indicate about where the water droplets came from?

_____

_____

_____

6 **Use Variables** What were the independent and dependent variables in this experiment? Which variables were controlled?

_____

_____

_____

_____

7 **Infer** Why do you think water droplets formed where they did?

_____

_____

_____

Explore More

What happened to the salt under the glass with water droplets? Plan and carry out an experiment that shows where the salt is.

_____

_____

_____

Open Inquiry

How quickly or slowly would water droplets form outside
a glass of solid ice? Think of your own question about
how quickly water droplets would form. Make a plan and
carry out an experiment to answer your question.

My question is:

_____

_____

How I can test it:

_____

_____

_____

_____

My results are:

_____

_____

_____

_____

Observing the Water Cycle

Materials

- 2 jars with tight-fitting lids
- water
- thermometer

① **Experiment** Fill two jars halfway with water. Use the thermometer to measure the temperature of the water in each jar. Record the temperatures in the data table below. Close each jar tightly with its lid.

	Water Temperature at Beginning	Water Temperature after 30 Minutes
Jar in sunny location		
Jar in dark location		

② Place one jar near a sunny window. Place the second jar in a cool, dark corner of the classroom.

③ Remove the lid from each jar. Use the thermometer to measure the temperature of the water in each jar. Record your results in the data table above. Which jar contains the warmer water?

_____

Name _____ Date _____

Observe Legume Roots

Materials

• legume root

• carrot root

• grass roots

1 Examine a legume plant. Clean off all dirt on the roots of the plant.

2 **Observe** Use a hand lens or microscope to examine the roots. What did you observe?

_____

_____

_____

3 Use a hand lens to examine a carrot root and grass roots. Compare these roots to the legume roots.

4 How are the legume roots similar to the other roots you observed? How are they different?

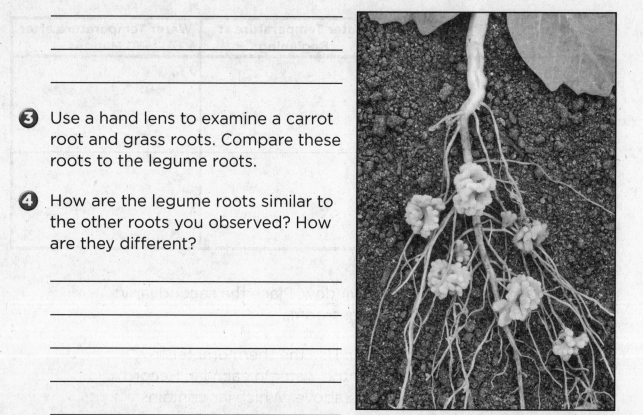

_____

_____

_____

_____

5 **Infer** Why are root nodules important in the nitrogen cycle?

_____

_____

_____

Name _____ Date _____

What happens when ecosystems change?

Materials

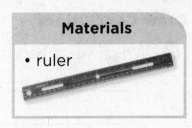

- ruler

Make a Prediction

Each year a tree grows wider as a new layer of xylem forms an annual ring. Scientists often use tree rings to study changes in ecosystems. How did this tree's ecosystem change? Make a prediction.

_____

_____

_____

_____

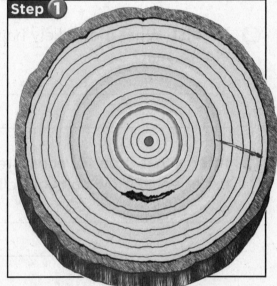

Step 1

Test Your Prediction

1 Count the number of tree rings on the tree ring diagram. How old was this tree?

_____

_____

2 **Measure** Use the ruler to measure the width of each tree ring. Record your measurements on another sheet of paper.

3 **Interpret Data** Use the information provided in the chart to interpret your tree ring data.

Tree Ring Data	
Type of Ring	**Event Affecting Tree**
thick ring	good growing conditions: warm, plenty of precipitation
narrow ring	poor growing conditions: cold, drought
dark scars	fire
long, light scars	insect infestation or disease

© Macmillan/McGraw-Hill

Draw Conclusions

4 Which years had the thickest rings? The narrowest rings?

_____

_____

_____

5 **Predict** What most likely happened to the tree in its eighth year?

_____

_____

6 **Infer** What types of ecosystem changes did this tree experience? How can you tell?

_____

_____

_____

Explore More

Have there been any fires, droughts, or floods in your community? Investigate using newspaper or Internet sources. Which parts of the environment have recovered better than others? Why?

_____

_____

_____

_____

© Macmillan/McGraw-Hill

Explore

Open Inquiry

What do you think would happen to the ecosystem
where this tree was found if all the trees were destroyed
by a fire? Think of your own question about how the
ecosystem would change. Make a plan and carry out
research to answer your question.

My question is:

_____

_____

How I can test it:

_____

_____

My results are:

_____

_____

_____

_____

_____

_____

Name _____ Date _____

How has the ecosystem changed around Mount St. Helens?

Materials
• magazines
• newspapers
• encyclopedias
• Internet

Make a Prediction

On May 18, 1980, Mount St. Helens, a volcano in Washington, erupted. How do you think the ecosystem changed during the eruption? How do you think the ecosystem has changed since the eruption? Make a prediction about what the ecosystem is like today.

_____

_____

Test Your Prediction

1 Research what the ecosystem surrounding Mount St. Helens was like before the eruption. Also research how the ecosystem changed during the eruption and at intervals following the eruption until the present date.

2 **Interpret Data** Describe the ecosystem surrounding Mount St. Helens before it erupted.

_____

_____

3 **Interpret Data** How did the ecosystem change during and immediately after the eruption?

_____

_____

4 **Interpret Data** How had the ecosystem changed after five years? 10 years? 20 years?

_____

_____

_____

© Macmillan/McGraw-Hill

Extinction Game

Wild Atlantic sturgeon are endangered because of overfishing and pollution.

Materials

- 20 pennies
- construction paper

1 Count out 20 pennies to represent a school of sturgeon.

2 **Make a Model** Tape a piece of construction paper to your desk. Divide it into 6 sections. Sections 1 and 3 represent death. Sections 2, 4, and 6 represent life. Section 5 represents a new offspring.

3 Toss all 20 pennies onto the paper.

4 Remove any pennies that land in sections 1 and 3. Add offspring for any pennies that land in section 5. Record the new number of sturgeon in a data table on another sheet of paper.

5 Play the game for 20 rounds (years). After each round, record the number of sturgeon.

6 **Communicate** Did your school of sturgeon become extinct? If so, how many years did it take?

_____

_____

_____

Name _____ Date _____

Interpret data

Ecosystem changes can affect organisms. Scientists estimate that once there were more than 500,000 bald eagles in America. But by the 1960s, there were less than 450 nesting pairs. What happened? Scientists discovered particles of an insecticide called DDT in the eagles' eggshells. The United States outlawed the use of DDT in 1972. Did that help bring eagles back from the edge of extinction? Scientists learned the answer to that question by collecting and **interpreting data**.

▶ **Learn It**

When you **interpret data**, you use information that has been gathered to answer questions or solve problems. It is much easier to interpret data that has been organized and placed on a table or graph. Tables and graphs allow you to quickly see similarities and differences in the data.

The table on the next page shows data gathered about bald eagle eggs. It lists the average number of eggs that hatched in the wild during a 16-year period. It also lists the levels of pesticide found in the eggs during that time.

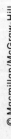

▶ **Try It**

Study the table, then **interpret data** to answer these questions:

1 In which year did the amount of pesticide in eggshells begin to decline? Why?

_____

_____

2 Did the amount of pesticide continue in a steady decline?

_____

_____

3 Does the data supply evidence that insecticide in eggs and the numbers of young hatched are related?

_____

_____

_____

Bald Eagle Hatching Data		
Year	**Average # Hatched**	**DDT in eggs parts/million**
1966	1.28	42
1967	0.75	68
1968	0.87	125
1969	0.82	119
1970	0.50	122
1971	0.55	108
1972	0.60	82
1973	0.70	74
1974	0.60	68
1975	0.81	59
1976	0.90	32
1977	0.93	12
1978	0.91	13
1979	0.98	14
1980	1.02	13

© Macmillan/McGraw-Hill

Hatched Eagle Eggs

Average Number of Hatched Eggs

0

1966 1967 1968 1969 1970 1971 1972 1973 1974 1975 1976 1977 1978 1979 1980 1981

Year

▶ **Apply It**

1 Now use the data from the table to make two line graphs: one to show the average number of eggs that hatched and one to show the insecticide in the eggs. Do your graphs make it easier to **interpret data**? Why or why not?

_____

_____

2 Lay one graph carefully on top of the other so the years across the bottom line up. Hold the pages up to the light. How would this help someone understand the relationship between the eagle eggs that hatched and the amount of insecticide in the eggs?

_____

_____

How are soils different?

Make a Prediction
The nutrient content of soils can vary greatly. The amounts of nutrients in soils can influence the types of organisms that can live in certain places. Which type of soil has more nutrients? Make a prediction.

_____

_____

Test Your Prediction

1 ⚠ **Be Careful.** Wear your goggles and apron. Place a spoonful of sand in the plastic cup.

2 **Observe** Add hydrogen peroxide to the sand, drop by drop. Hydrogen peroxide is a chemical that bubbles when it reacts with nutrients.

3 **Communicate** Record the number of drops it takes until the sample starts bubbling. Number of drops:

Sand:_____

4 Repeat steps 1 to 3 using soil instead of sand. Record your data.

Soil:_____

Materials
- goggles
- apron
- plastic spoons
- sand
- plastic cups
- hydrogen peroxide
- dropper
- soil

Step **1**

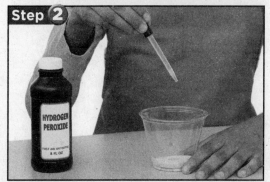
Step **2**

Draw Conclusions

5 Which sample had more nutrients—sand or soil? Explain.

_____

_____

_____

6 **Predict** Which would probably be better for growing plants—sand or soil? Explain.

_____

_____

7 **Infer** How might you classify the sand and soil—high in nutrients or low in nutrients?

_____

_____

Explore More

Collect other types of soil and test their nutrient levels using hydrogen peroxide. Which soil had the greatest amount of nutrients?

_____

_____

_____

© Macmillan/McGraw-Hill

Open Inquiry

What would happen to a plant if it were planted in sandy soil? Think of your own question about what a plant needs to survive. Make a plan and carry out an experiment to answer your question.

My question is:

_____

_____

How I can test it:

_____

_____

My results are:

_____

_____

_____

_____

What nutrients are in packaged soils?

Make a Prediction

Garden centers often sell a variety of soils in bags. These include potting soils, top soils, and composted soils. Which of these materials do you think has the most nutrients? Make a prediction.

_____

_____

Materials
• potting soil
• hydrogen peroxide
• cups
• spoon
• dropper
• apron
• goggles

Test Your Prediction

1 ⚠ **Be Careful!** Put on goggles and an apron Handle the hydrogen peroxide carefully. Do not spill it on your clothing, skin, or the work area.

2 Label as many clean plastic cups as the number of soil samples. Place a spoonful of soil into each labeled cup.

3 **Observe** Add hydrogen peroxide to each cup, drop by drop, until the soil bubbles. Count the drops carefully as you add them. Record your results for each type of soil in a table on a separate sheet of paper.

Draw Conclusions

4 **Draw Conclusions** Which soil sample had the most nutrients?

_____

_____

Name _____ Date _____

Compare Leaves

Materials
• succulent leaf
• broad leaf

1. **Observe** Compare the broad leaf and the succulent leaf. Which leaf is thicker? Which leaf would be better for catching sunlight? Record your observations.

   _____

   _____

   _____

2. Break open each leaf by tearing off a piece of it. Which leaf holds more water?

   _____

   _____

3. **Infer** Think about the characteristics of each leaf. Which leaf might have come from a forest? Which leaf might have come from a desert? Explain.

   _____

   _____

   _____

   _____

Name _____ Date _____

How does the ocean get salty?

Purpose
To make a model that shows how ocean water becomes salty.

Procedure

1 Measure In the plastic cup, mix 2 tablespoons of salt and a few drops of food coloring. Use the spoon to stir until it's well mixed.

2 Pour 2 cups of soil into one side of the shallow baking pan.

3 Mix the salt with the soil in the pan.

4 Tip the pan so the side with the mixture in it is slightly off the table. Try not to knock any of the mixture to the other side.

5 As you hold the pan slightly off the table, slowly pour some water onto the mixture.

6 Observe Note the color of the water when it reaches the other side of the pan.

Materials

- plastic cup
- salt
- blue food coloring
- plastic spoon
- soil
- baking pan
- container of water

Step **1**

Step **5**

© Macmillan/McGraw-Hill

Draw Conclusions

7 How does the color of the water compare to the color of the dyed salt?

_____

_____

8 **Infer** How does this model resemble what happens as fresh water flows to the ocean?

_____

_____

Explore More

Are some oceans saltier than others? Research Earth's oceans to find out if some have more salt than others. Write a report that explains how some oceans become saltier than others.

_____

_____

_____

_____

© Macmillan/McGraw-Hill

Open Inquiry

Is salt water found on Earth in places other than the oceans and seas? Think of your own question about where salt water is located on Earth. Make a plan and carry out research to answer your question.

My question is:

_____

_____

How I can test it:

_____

_____

My results are:

_____

_____

_____

_____

What is ocean water made of?

Materials

• reference books

Purpose

Ocean water is salty because it contains the element sodium. Ocean water also contains other chemicals and elements. What other substances are found in ocean water?

Procedure

1 Use books, magazines, encylopedias, or the Internet to find out what ocean water is made of.

2 Record your findings in the data table below.

Element/Mineral	Amount
Sodium	

Draw Conclusions

3 **Analyze Data** What elements or minerals have the highest percentages in ocean water?

_____

_____

_____

Name _____   Date _____

Salt Water vs. Fresh Water

Materials

- salt water
- fresh water
- 2 carnations
- 2 plastic cups

1. Fill a cup with fresh water. Fill a cup with salt water. Label each cup. Place flowers in each cup.

2. **Observe** Examine each flower after 2 hours.

3. **Communicate** Did you see any changes to either flower? Explain your observations.

_____

_____

_____

4. **Infer** What caused the changes in the flower in the salt water?

_____

_____

_____

_____

5. Is salt water a good ecosystem for any living things?

_____

_____

_____

_____

Name _____ Date _____

What are Earth's features?

Purpose

To examine and classify Earth's features.

Procedure

① **Observe** Examine the photos.

② List the features of Earth's surface that you can identify in these photos.

_____

_____

_____

_____

_____

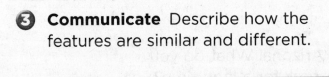

③ **Communicate** Describe how the features are similar and different.

_____

_____

_____

_____

_____

Chapter 5 • Our Dynamic Earth
Activity Lab Book

Name _____ Date _____

Draw Conclusions

4 **Classify** Identify groups into which you could sort Earth's features.

_____

_____

_____

_____

5 **Infer** What processes might have produced one or more of the features you have identified?

_____

_____

_____

_____

Explore More

Find photos of the Grand Canyon in Arizona. What do you think happens when water runs over rock for a long time? Form a hypothesis about how water was involved in forming the canyon. Design an experiment that would test your prediction.

_____

_____

_____

_____

_____

Open Inquiry

Suggest a specific landform, either one in your area or a well-known one such as the Grand Canyon. Hypothesize about how it might have been formed.

▶ My question is:

_____

_____

▶ How I can test it:

_____

_____

▶ My results are:

_____

_____

_____

Name _____ Date _____

What features does Earth have here?

Materials
- paper
- pencil

Purpose
Identify local landforms

Procedure

1 **Observe** Look out the window. Notice the features of Earth's surface, such as hills, streams, or deserts.

2 What features do you see?

_____

_____

_____

3 **Classify** Divide your list of features into groups. For each group, list other features that would fit in the same group.

_____

_____

_____

Mapping the Ocean Floor

Materials
- soft clay
- one sandwich container with lid
- dowel for probe
- ruler

1 Place soft clay in the bottom of a container to form features of the ocean floor.

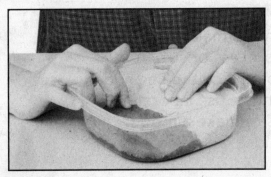

2 Cover your container.

3 Exchange your model with another classmate.

4 **Measure** Gently drop a probe into each hole and measure how much of the probe sticks out of the hole. Record the depth of each square of the grid.

Coordinates	Depth

5 **Interpret Data** Use your probe measurements to figure out the height of the features. Then draw and label them.

6 Remove the top of the container and compare your drawing to the ocean features.

_____

_____

Make a Model

Models show the basic features of a structure or process. When scientists **make a model**, they are simplifying a process or structure that would otherwise be difficult to see and understand. Many scientists use laboratory materials or computers to **make a model** so they can explain an idea, an object, or an event.

Mt. Rainier

▶ **Learn It**

What is the relationship between elevation and contour lines? Mapmakers use elevation measurements that are made by surveyors to make contour lines on maps. The surveyors use various instruments that can accurately measure distances. Telescopes, aerial photographs, and satellite images help them to get those elevation measurements.

How can you understand the relationship between elevation and contour lines without those instruments? **Make a model** of a mountain and use it to make a topographic map to help you understand the relationship between contour lines and elevation.

Mt. Hood

▶ **Try It**

1 **Make a Model** Form the clay into a mountain shape.

2 Using the pencil, poke a hole through the center of your mountain.

3 Measure the height of your mountain in centimeters.

4 Using the dental floss, cut a 1-centimeter slice off the top of the mountain.

5 Place the slice of mountain onto a piece of graph paper. Trace the edges of the slice and mark the location of the hole in the middle. Then put the slice to the side.

6 Cut the next 1-centimeter slice of mountain. Place it on the graph paper so the hole in the middle of the slice lines up with the spot you marked on the graph paper from the previous slice. Then trace the edges of this slice.

7 Cut, line up, and trace the rest of the slices of mountain. When you are finished with the bottom slice, put the clay mountain back together.

Materials

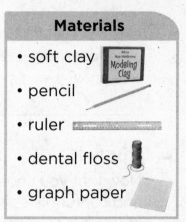

- soft clay
- pencil
- ruler
- dental floss
- graph paper

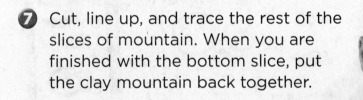

© Macmillan/McGraw-Hill

Name _____ Date _____

8 Is there a point on your topographic map where the lines are closer together? What does the mountain look like in this area?

_____

▶ **Apply It**

Now that you understand the relationship between elevation and contour lines, you can apply your understanding when reading topographic maps. The topographic map below shows a student's neighborhood with the contour lines measured in feet. Look at the topographic map below and answer the following questions.

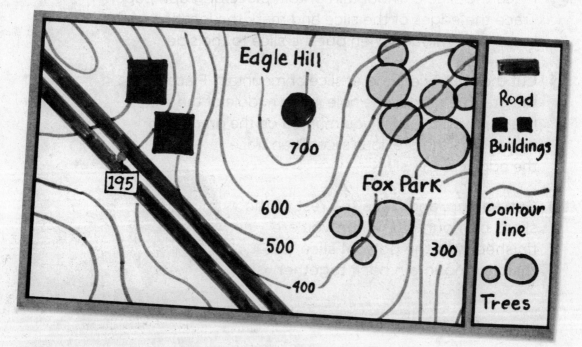

▶ What is the lowest elevation on the map?

_____

▶ What is the highest elevation on the map?

_____

▶ The dark circle represents a school. What is its elevation?

_____

How does a mountain range form?

Purpose
To model one way a mountain range can form.

Procedure

1 Lay the sheet of newspaper on your desk. In the middle of the sheet of newspaper, place one sheet of construction paper about 5 centimeters over the other sheet.

2 Spread sand over the overlapped pieces of construction paper. Then smooth the sand with your hand.

3 **Make a Model** Hold down the farther-away piece of construction paper with one hand. With your other hand, slowly push the closer sheet of construction paper underneath the other sheet.

4 **Observe** What happens to the sand?

_____

_____

_____

Materials

- 1 sheet of newspaper
- 2 sheets of construction paper
- sand

Name _____ Date _____

Draw Conclusions

5 **Infer** In this model, the mountain range that formed was straight. How could you change the model to make the mountain range curve?

_____

_____

6 What would you expect to happen if you continued to push on the closer sheet?

_____

_____

Explore More

In this model of mountain range formation, you pushed one sheet of paper underneath another. What landforms would you model if you pulled the sheets of paper in other directions?

_____

_____

_____

_____

_____

© Macmillan/McGraw-Hill

Open Inquiry

What would happen with the collision of two separate landmasses?

▶ My question is:

_____

_____

_____

▶ How I can test it:

_____

_____

_____

▶ My results are:

_____

_____

_____

How do some mountains form?

Make a Prediction

You are going to slide two slabs of modeling clay into each other. Predict what will happen to them.

_____

_____

Materials

- modeling clay
- ruler
- paper
- scissors

Test Your Prediction

1 Roll out two slabs of clay 15 centimeters square and 1.5 centimeters thick.

2 Cut two pieces of paper 16 centimeters square. Place a slab on each one.

3 Place the slabs about 16 centimeters apart. Cover half the clay of each slab with your hand. Leave the half closest to the other slab uncovered.

4 Slowly slide your hands toward each other. Do not stop when the slabs touch each other.

5 **Observe** What happened to the clay?

_____

_____

Draw Conclusions

6 Was your hypothesis correct?

_____

_____

7 **Infer** This procedure was a model of how some mountains form. What does it tell you about the real-life process it modeled?

_____

© Macmillan/McGraw-Hill

Spread of the Ocean Floor

Materials

- two stacks of books
- one sheet of paper
- ruler
- two colored markers (optional)

1. Place two stacks of books next to each other with a narrow space between them.

2. Fold a sheet of paper and slip it into the space between the books.

3. **Measure** Using a ruler, measure 3 cm of the paper. Fold that much paper down onto the books. Color this part of the sheet purple.

4. **Make a Model** Move the sheets of paper up another 3 cm. Color the newly exposed areas of both sheets with another color. Repeat this step until you run out of paper.

5. If the two sheets of paper represent the spreading sea floor, which color represents the youngest rock?

_____

_____

_____

© Macmillan/McGraw-Hill

Name _____ Date _____

How can you predict when a volcano will erupt?

Purpose
To build a tiltmeter, or an instrument that can measure changes in the slope of a volcano.

Procedure

1. ⚠ **Be Careful.** Using the tip of a pencil, punch a small hole into the side of a foam cup about 2.5 cm from the bottom of the cup. Push the coffee stirrer into the hole so it fits tightly.

2. Punch a hole in the second cup. Push the other end of the coffee stirrer into that hole.

3. Mix 3 drops of food coloring into a container of water.

4. Put the cups and stirrer into a baking pan. Then pour each cup about half full of the dyed water.

5. **Observe** Carefully tilt one end of the pan. What happens to the level of the water in the cups?

_____

_____

_____

Materials

- pencil
- 2 foam cups
- coffee stirrer
- food coloring
- container of water
- baking pan

Step 1

Step 4

© Macmillan/McGraw-Hill

Draw Conclusions

6 **Experiment** Raise the tiltmeter to different heights and measuring the change in the height of the water. How well does the tiltmeter record changes in height?

_____

_____

_____

7 **Infer** What might it mean if a tiltmeter measured an increase in the steepness of a volcano's slope?

_____

_____

_____

Explore More

How would the tiltmeter work with different-sized cups or different lengths of straw? Decide on a variable you want to change. Design an experiment to test how well the model with the changed variable measures tilt.

_____

_____

_____

Name _____ Date _____

Open Inquiry

What is something else you could measure with a tiltmeter?
Design an experiment with a tiltmeter.

▶ My question is:

_____

_____

_____

▶ How I can test it:

_____

_____

_____

▶ My results are:

_____

_____

_____

How can you measure slope?

Materials

- 2 rulers
- tape
- string
- 10 g weight or similar object
- protractor

Purpose
Measure the slope of a surface

Procedure

1 Tape the ends of the rulers together so that they form a right angle.

2 Tie the weight to one end of the string. Tape the other end of the string to the place where the rulers are attached.

3 Hold the rulers upright with the ends resting on a flat surface. Hold the protractor between the two rulers so that the 0° mark lines up with the string. Have your partner tape the protractor to the rulers in that position.

4 Lean a book or other long, flat object against the wall.

5 **Measure** Place the ends of both rulers on the leaning book. Read the angle from the protractor that is shown by the string.

_____

Draw Conclusions

6 What would your tiltmeter read if you placed the ends of both rulers on the wall?

_____

Name _____ Date _____

Types of Volcanoes

Materials

- modeling clay
- paint
- glue
- safety scissors
- construction paper
- paper towel tubes
- index cards

1 **Make a Model** Use materials of your choice to make models of the three kinds of volcanic mountains.

2 **Communicate** Write descriptions of each type of volcano and how it was formed. Place the description next to each model.

3 **Compare** How are the layers of the volcano models you made similar to or different from the layers of the volcanic mountains?

_____

_____

Name _____   Date _____

Structured Inquiry

How do volcanoes form islands?

Form a Hypothesis

When tectonic plates move over hot spots at different speeds, what do the islands that form look like? Write your answer as a hypothesis in the form "If one tectonic plate is moving faster than another over a hot spot, then . . ."

_____

_____

Test Your Hypothesis

Materials

- plaster of Paris
- measuring cup
- spoon
- funnel
- squeeze bottle
- large container
- 1 piece of cardboard
- tray

① **Measure** △ **Be Careful.** Wear goggles. Place 750 mL of plaster of Paris into a large container. Add 250 mL of water and stir the mixture until a thin paste forms.

② **Make a Model** Pour this mixture into the squeeze bottle. This mixture represents magma. The nozzle of the bottle represents the hot spot.

Step ①

③ **Make a Model** Place the tip of the bottle at one end of the hole in the cardboard. The cardboard represents the tectonic plate.

④ Gently squeeze the bottle until the lava starts to flow up through the hot spot. Continue to squeeze the bottle as you pull the piece of cardboard towards you. What happens?

Step ②

_____

_____

_____

_____

© Macmillan/McGraw-Hill

5 Refill the bottle with the plaster of Paris and water mixture. Place the tip of the bottle into the end of a second hole in the cardboard. Slowly pull the piece of cardboard towards you as you squeeze the bottle. What happens?

_____

_____

Draw Conclusions

6 Compare what happened as a result of steps 4 and 5. Do the results appear different? If so, why?

_____

_____

_____

_____

_____

7 **Infer** How do volcanic islands appear if the tectonic plate is moving slowly over a hot spot?

_____

_____

8 **Infer** How do volcanic islands appear if the tectonic plate is moving rapidly over a hot spot?

_____

_____

Guided Inquiry

How do eruptions of different types of lava affect the height of a volcano?

Form a Hypothesis

You know that the shapes of volcanoes are different when they form from lava of different thickness. How will thicker or thinner lava affect the height of a volcano? Write your answer as a hypothesis in the form "If thicker lava is used to form a volcano, then . . ."

_____

_____

_____

Test Your Hypothesis

Design an experiment to investigate effect of eruptions with different types of lava on the height of volcanoes. Write out the materials you need and the steps you will follow. Record your results and observations.

_____

_____

_____

_____

Draw Conclusions

Did your results support your hypothesis? Why or why not? Present your results to your classmates.

_____

_____

Open Inquiry

Does lava with bubbles of gas in it move differently than lava without bubbles? What happens to the area around a volcano as the volcano erupts? Design an experiment to answer your question. Your experiment must be organized to test only one variable. Keep careful notes as you do your experiment so another group could repeat the experiment by following your instructions.

▶ My question is:

_____

_____

_____

▶ How I can test it:

_____

_____

_____

▶ My answer is:

_____

_____

_____

How does ground move during an earthquake?

© Macmillan/McGraw-Hill

Materials

- cut pieces of foam
- pan
- soil
- wooden block

Purpose

To model the movement of the ground during an earthquake.

Procedure

1 Place 2 pieces of foam in a pan so the cut surfaces touch each other.

2 Cover the foam with soil and smooth the soil over both pieces of foam.

3 Pull about 5 centimeters of the pan off of the edge of the table.

4 **Observe** ⚠ **Be Careful.** Gently tap the bottom of the pan with a block. What happened to the blocks and the soil?

_____

_____

Step **1**

5 What happens as you continue to tap the pan?

_____

_____

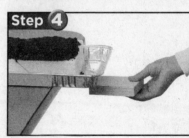

Step **4**

Name _____ Date _____

Draw Conclusions

6 **Infer** What would happen if you tapped the pan harder?

_____

_____

7 What do the foam blocks and the cut between the blocks represent?

_____

_____

Explore More

In this model, the cut between the foam blocks has a certain angle. How do you think the model would work if the blocks were cut at a different angle? Form a hypothesis about which angle will cause more dirt to fall. Make a model and test your hypothesis.

_____

_____

Name _____ Date _____

Inquiry Open

Design a new model that will reflect what happens in an earthquake more accurately.

▶ My question is:

_____

_____

▶ How I can test it:

_____

_____

▶ My results are:

_____

_____

_____

Name _____ Date _____

How does the ground move during an earthquake?

Materials
• 2 board erasers
• plastic wrap
• newspaper
• potting soil

Purpose
Model an earthquake

Procedure

1. Wrap the board erasers in plastic wrap to keep them clean.

2. Cover a small desk or table with newspaper.

3. Place the erasers on the small desk or table. Slide them close together.

4. Sprinkle soil over the two erasers until you cannot see the edge where they meet.

5. Rock the desk back and forth.

6. **Observe** What happened to the erasers and soil?

_____

_____

Draw Conclusions

7. What does the line where the erasers touched represent?

_____

_____

© Macmillan/McGraw-Hill

Modeling P and S Waves

Materials
• Coiled toy

① While your partner holds one end of a coiled toy, stretch the toy until it is fully extended.

② **Make a Model** Using your thumb, pluck one of the coils. What happens? As you move the toy, draw or describe the motion that one coil makes.

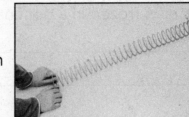

_____

_____

_____

③ Which type of wave does this model represent?

_____

④ **Make a Model** Stretch the coiled toy again. Move your end up and down. What happens? As you move the toy, draw or describe the motion that one coil makes.

_____

_____

⑤ Which type of wave does this model represent?

_____

Name _____ Date _____

How does ice break up rocks?

Form a Hypothesis

Which takes up more space, liquid water or frozen water? Write your answer as a hypothesis in the form "If water is frozen in a confined space, then . . ."

_____

_____

Materials

- metric ruler
- marker
- two identical plastic containers with caps
- food coloring
- water

Test Your Hypothesis

1 Using the marker and ruler, mark 15 centimeters on each container.

2 Mix 5 drops of food coloring into the water.

3 Fill both containers with water until the water reaches the 10-centimeter mark.

4 Put caps on both containers. Place one container in a freezer. Leave the other container at room temperature.

Step 1

5 When the water in the container in the freezer has completely frozen, remove the container.

6 **Observe** Is there a change in the height of the water in either container? Is there a change in the shape of either container?

_____

_____

Step 3

Draw Conclusions

7 **Interpret Data** What happens to the amount of space water takes up when it freezes?

_____

8 **Infer** What do the results of your experiment indicate about what happens when water freezes in a crack in a rock?

_____

_____

_____

Explore More

Other processes can change the surfaces of rocks. Observe the sidewalks in your neighborhood and pay special attention to cracks or changes in their surfaces. What might have caused these changes?

_____

_____

_____

Name _____ Date _____

Open Inquiry

What happens when other substances freeze? Design an experiment comparing another substance to water when it freezes.

▶ My question is:

_____

_____

▶ How I can test it:

_____

_____

_____

▶ My results are:

_____

_____

_____

What happens to water when it freezes?

Make a Prediction

Your teacher has poured equal amounts of water into two cups and frozen one of them. When you look at the cups, what do you think you will observe?

_____

_____

_____

Test Your Prediction

1 **Observe** Look at the two cups. What do you see?

_____

_____

2 **Infer** What happened to the water as it froze?

_____

_____

Draw Conclusions

3 Was your prediction correct?

_____

Name _____ Date _____

Rate of Erosion

Materials

• two identical baking pans

• two wooden blocks

• watering can

• measuring cup

• water

• dirt

① **Form a Hypothesis** How does the speed of running water affect how fast soil erodes? Write your answer in the form of a hypothesis.

_____

_____

② **Make a Model** Place dirt in two identical baking pans so the dirt is at the same level in each pan.

③ Place a wooden block under each pan.

④ Fill a watering can with a sprinkler head with 2 cups of water. Slowly pour the water into the pan. Record your observations.

_____

_____

_____

⑤ Remove the sprinkler head and fill the watering can with 2 cups of water. Pour the water slowly into the pan. Record your observations.

_____

_____

_____

⑥ **Draw Conclusions** Do your results support your hypothesis?

_____

_____

_____

What are the properties of minerals?

Purpose

To observe the properties of minerals.

Procedure

1 Use the clear tape and the marker to label each mineral with a different sample number.

2 Use the chart below, or make one   on another sheet of paper.

Materials

• mineral samples

• clear tape

• marker

• porcelain tile

• copper penny

• steel file

Sample Number	Mineral	Color	Shine (yes/no)	Streak	Scratch	Other
1						
2						
3						
4						

Name _____   Date _____

3 **Observe** Fill in the columns of the chart for *color* and *shine* (like a metal).

4 **Observe** Rub the mineral across the porcelain tile. Record the color that you see on the tile.

5 **Observe** △ **Be Careful.** Scratch the mineral on a copper penny and a steel file. Record whether the mineral scratches the penny or the file.

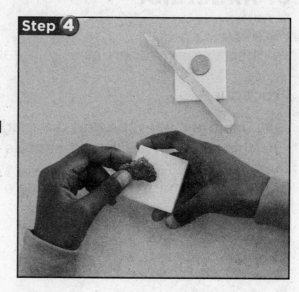
Step 4

Draw Conclusions

6 **Infer** Examine your data. What can you say about the properties of different minerals?

_____

_____

_____

7 How do the properties of minerals help you classify minerals?

_____

_____

_____

Explore More

Using reference sources, identify these minerals. Then label and display them.

Open Inquiry

Design an experiment to see if two minerals are the same or different.

My question is:

_____

_____

How can I test it?

_____

_____

_____

My results are:

_____

_____

_____

Name _____ Date _____

Where are minerals found?

Materials
• rocks
• hand lens
• penny
• steel
• porcelain tile

Purpose
Show that rocks are made up of minerals.

Procedure

1 Obtain some rocks from your teacher.

2 **Observe** Use the hand lens to look at the minerals you can see in the rock. Describe what you see.

_____

_____

3 **Analyze** How might you identify the minerals you see?

_____

_____

_____

4 **Apply** Identify as many minerals as you can in the rocks. In addition to the hand lens, you can use the tests you used in the Explore activity.

_____

_____

5 What is the relationship between rocks and minerals?

_____

_____

Crystal Shapes

© Macmillan/McGraw-Hill

Materials

- goggles
- kitchen mitt
- plastic spoon
- cup of very warm water
- string
- sugar
- sugar cubes
- pencil

1 ⚠ **Be Careful.** Wear goggles. Use a kitchen mitt if you need to hold or move the cup. Don't touch the very warm water.

2 Using a plastic spoon, slowly add small amounts of sugar to a cup of very warm water. Continue to add sugar and stir until you can see sugar in the water.

3 Tie one end of a 15-cm length of string around a sugar cube. Tie the other end to a pencil. Place the pencil across the cup so that the cube hangs in the very warm water without touching the sides or bottom.

4 **Observe** Watch the experimental setup for several days.

5 **Communicate** Describe what you observed in the cup.

_____

Name _____ Date _____

Classify

As you just read, rocks are naturally formed solids made up of one or more minerals. Each mineral adds its own properties to a rock. There are billions of different rocks on Earth. Scientists group, or **classify**, rocks into three groups based on the way they form. In order to determine how they form, scientists observe the properties of the rocks. These properties include color, weight, texture, and whether the rocks float or sink.

▶ **Learn It**

When you **classify**, you group objects that share properties. You need to compare and contrast the objects in order to find out what properties they share. Remember, to compare you look at how things are alike, while to contrast you look at how they are different.

Classifying is a useful tool for organizing and analyzing. It can help you understand why things belong in the same groups and how some things can belong to several different groups. It is important to keep notes as you **classify**. Your notes can help you figure out how to classify other things.

Use with **Lesson 1**
Minerals and Rocks

© Macmillan/McGraw-Hill

Name _____ Date _____

▶ **Try It**

① Make a table or use the one below. List the properties that you want to look for in the first column.

Materials
- 8 different rocks
- water
- small bowl

Classifying Rocks by Properties								
	#1	#2	#3	#4	#5	#6	#7	#8
color: dark								
color: light								
several colors								
heavy								
light								
rough								
smooth								
sharp								
has holes								
has layers								
floats								
sinks								

② Examine the first rock carefully.

③ **Classify** Mark an X in the appropriate box if this rock can be classified by the property listed in the rows.

④ Fill the bowl with water. Place the rock in the bowl to test whether or not the rock floats.

⑤ Repeat using the remaining rocks.

Name _____ Date _____

► **Apply It**

Now that you have classified rocks by their properties, look for rocks on the ground and in buildings. Make a table that lists all of the rocks that you have seen. List the properties by which these rocks can be classified. Finally, mark the chart to show which rocks have the same properties and can be classified together.

6 How many rocks would you classify as smooth?

_____

7 How many rocks would you classify as having layers?

_____

8 Which property is shared by the most rocks?

_____

9 Decide whether each rock is an igneous, sedimentary, or metamorphic rock.

Name _____ Date _____

What is in soil?

Purpose
To examine the contents of a soil sample.

Procedure

1 **Observe** Use the toothpicks and hand lens to separate the contents of the soil sample.

2 **Record Data** Identify and list the different materials in the soil sample.

- soil sample
- hand lens
- tooth picks

What's in the soil

© Macmillan/McGraw-Hill

Draw Conclusions

3 **Classify** Does your soil sample contain nonliving things? What about once-living things?

_____

_____

4 Based on your observations and data, what are the contents of soil?

_____

_____

Explore More

Collect and examine samples of soil from various places in your neighborhood. How do the contents of these samples compare with the one you studied in this activity? Do the additional samples change the conclusion you drew about the contents of soil?

_____

_____

_____

© Macmillan/McGraw-Hill

Open Inquiry

Design an experiment to see how decayed plant and animal parts contribute to plant growth.

My question is:

_____

_____

How can I test it?

_____

_____

My results are:

_____

_____

_____

_____

Name _____ Date _____

How is potting soil different from other soil?

Purpose
Compare and contrast different types of soil.

Procedure

1 Obtain some potting soil and some other soil from your teacher. Place each type of soil on a separate piece of paper.

2 **Observe** Use the toothpicks and the hand lens to examine the potting soil. What do you see?

_____

_____

3 **Observe** Use the toothpicks and the hand lens to examine the other soil. What do you see in this soil?

_____

_____

4 **Compare** What do the two types of soil have in common?

_____

_____

5 **Contrast** How are the two types of soil different?

_____

_____

6 How do you think the soils became so different?

_____

_____

© Macmillan/McGraw-Hill

Soil Soaks Up Water

Materials

- topsoil
- sand
- bowl
- pen
- three cups
- large measuring cup
- small measuring cup
- water

1 In a bowl, measure out topsoil and sand to make a soil mixture that you predict will hold water well.

2 ⚠ **Be Careful.** Using the point of a pen, punch an equal number of small holes in the bottom of three cups.

3 Fill one cup with topsoil, one cup with sand, and one cup with your mixture.

4 **Experiment** While holding the cup with the topsoil over a large measuring cup, pour 100 mL of water into the cup with soil. Allow the water to drain through the cup for 5 minutes.

5 Measure the water that passed through the soil.

Water Measurements		
topsoil	sand	my mixture

6 Repeat steps 4 and 5 with the sand and with your mixture.

7 Calculate the amount of water that the soil soaked up.

_____

8 **Interpret Data** Which type of soil holds the most water?

_____

Name _____ Date _____

Structured Inquiry

Which soil is better for plant growth?

Form a Hypothesis

Different types of soil are made of different materials. Sand is a type of soil made from small pieces of rocks. Potting soil is made from bits of sticks and leaves. How fast will grass seeds grow in potting soil compared to sand? Write your answer as a hypothesis in the form "If grass seeds are planted in potting soil and in sand, then . . ."

Materials

- 2 pans
- 2 measuring cups with water
- sand
- potting soil
- grass seeds
- ruler

_____

_____

Test Your Hypothesis

❶ Fill one pan with potting soil until the soil is 1 inch deep. Fill the other pan with sand until the sand is 1 inch deep.

Step ❶

❷ Evenly scatter grass seeds over each pan.

Step ❷

© Macmillan/McGraw-Hill

③ Place the pans in the sunlight.

④ Every other day, pour the same amount of water on the seeds in both pans.

⑤ **Observe** What do the pans look like after three days? After one week?

_____

_____

Draw Conclusions

⑥ Why is it important to make sure the pans get the same amount of light and water?

_____

_____

_____

_____

⑦ **Infer** What differences between the potting soil and the sand may have affected the plant growth?

_____

_____

_____

© Macmillan/McGraw-Hill

Name _____ Date _____

Guided Inquiry

What effect does pollution have on plants?

Form a Hypothesis

You now know the type of soil in which plants will grow faster. How fast will plants grow in polluted soil? Write your answer as a hypothesis in the form "If grass seeds are planted in soil and polluted soil, then . . ."

_____

_____

_____

Test Your Hypothesis

Design an experiment to investigate how fast plants will grow in soil compared to polluted soil. Write out the materials you need and the steps you will follow. Record your results and observations in the chart on the next page.

Materials I Need

© Macmillan/McGraw-Hill

Effects of Pollution on Growth	
soil	polluted soil

Draw Conclusions

Did your results support your hypothesis? Why or why not? Present your results to your classmates.

_____

Open Inquiry

How efficient are conservation methods that slow down the flow of water over soil? Think of a question and design an experiment to answer it. Your experiment must be organized to test only one variable. Keep careful notes as you do your experiment so another group could repeat the experiment by following your instructions.

My hypothesis is:

_____

_____

How can I test it?

_____

_____

My conclusions are:

_____

Name _____ Date _____

How can wind move objects?

Form a Hypothesis

How many paper clips do you think you can move with your breath using a windmill? Write your answer as a hypothesis in the form "If the speed of the wind against a windmill blade increases, then . . ."

_____

_____

_____

Materials

- 8 cm by 15 cm strip of paper
- new pencil
- tape
- four 8 cm by 5 cm strips of paper
- paper clip
- string

Test Your Hypothesis

1 Wrap the 8 cm by 15 cm strip of paper around the pencil. Have a partner tape the edges of the paper together to form a tube.

2 Tape the 5 cm side of the 8 cm by 5 cm strips to the tube of paper near one end of the tube to make blades for the windmill. Space the strips so they are equally far apart.

3 Tie one paper clip to the string. Tape the other end of the string to the paper tube.

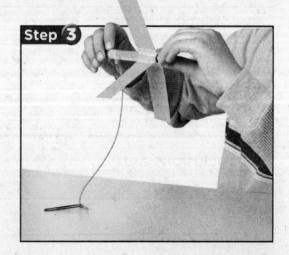

Step **1**

Step **3**

© Macmillan/McGraw-Hill

4 Hold the ends of the pencil and blow on the paper strips. What happens to the paper clip?

_____

_____

5 **Experiment** Now attach more paper clips to that one. How many paper clips did you add before your breath can no longer lift them?

_____

_____

_____

Draw Conclusions

6 How is the energy from your breath used to raise the paper clip?

_____

_____

7 **Infer** If you used larger rectangles for windmill blades, what do you think would happen to the number of paper clips you could lift?

_____

_____

Name _____ Date _____

Explore More

What result do you think you would get with different-shape blades? Think of a shape to test and come up with a design. Then experiment to find out whether your shape works better than a rectangle.

_____

_____

_____

Open Inquiry

When you blow softly on the blades, or with a little or a lot more force, how does it affect the work done? Think of your own question about the amount of wind needed to do work. Make a plan and carry out an experiment to answer your question.

My question is:

_____

_____

How can I test it?

_____

_____

_____

My results are:

_____

_____

_____

How much work can wind do?

Purpose

Determine how much work wind can do.

Procedure

Materials
• pinwheel
• electric fan
• 8 cm string
• hole punch
• paper clips
• safety goggles

1 Use the hole punch to punch a hole near the tip of one of the blades of the pinwheel.

2 Tie one end of the string through the hole. Tie a paper clip to the other end of the string.

3 ⚠ **Be Careful.** Put on safety goggles in case a paperclip comes loose and be careful with the fan and the pinwheel when the paper clip moves.

4 Have your teacher plug in the electric fan. Turn the electric fan on the slowest speed. Do not allow any object to come near the moving blades.

5 **Experiment** Hold the pinwheel in the air from the fan, away from any object that the paper clip might hit. Does the pinwheel raise the paper clip?

_____

6 **Observe** Add another paper clip to the first paper clip and repeat Step 5. What do you observe?

_____

7 **Collect Data** Repeat Step 6 until the pinwheel can no longer raise the paper clips. What is the greatest number of paper clips the pinwheel raised?

_____

8 **Conclude** Why do you think an electric fan might give you better results than blowing on the pinwheel?

_____

Name _____ Date _____

Half-Life of a Penny

Materials

• pennies

• boxes with lids

1. **Record Data** Count the total number of pennies that you were given.

_____

2. Place all of the pennies in a box so they are heads up.

3. **Experiment** Close the box and shake it to mix the pennies.

4. Open the box and remove all the pennies that have turned tails up. Set them aside.

5. Record the number of pennies remaining in the box.

_____

6. Repeat steps 2-4 until one or no pennies remain in the box.

7. How many pennies were removed after each shake?

_____

_____

8. What is the "half-life" of a penny?

_____

Use with **Lesson 3**
Fossils and Energy

© Macmillan/McGraw-Hill

Name _____ Date _____

How much fresh water do you use?

Materials

- sink
- container
- measuring cup

Make a Prediction

How much water do you use in a day for a particular activity, such as brushing your teeth or washing your hands?

_____

_____

Test Your Prediction

1 Put the container in the sink.

2 Turn the water on and pretend to brush your teeth or wash your hands. Run the water as long as you would if you were really doing that activity. Once you are done, turn the water off.

Step 2

3 **Measure** Using the measuring cup, scoop water out of the container into the sink. Keep track of each cup that you pour so you can estimate the total amount of fresh water you use for that activity.

_____

_____

Name _____ Date _____

Draw Conclusions

4 **Use Numbers** On a chart, figure out how many gallons of fresh water you use for the activity in a week, a month, and a year. One gallon = 16 cups.

Activity:	

5 **Communicate** Discuss how much water you used with your classmates. Exchange data for the amount of water you used for your chosen activity. Whose use of water was closest to their prediction?

6 Design and complete tables or graphs to display the results of all of the data collected by the other students.

Explore More
Think of a way you can reduce the amount of water that you used. Predict how much water you can save. Redo the activity you chose using your new idea. Were you able to save water? Discuss your idea and its result with your classmates.

_____

_____

Open Inquiry
How can you save water in the kitchen? Think of your own question about what kitchen activities use water and how you can use less water. Make a plan and carry out an experiment to answer your question.

My question is:

_____

_____

_____

How can I test it?

_____

_____

My results are:

_____

_____

Name _____ Date _____

How can you save shower water?

Materials
- shower
- bucket
- clock or stopwatch

Purpose

Measure the amount of water used during a shower and determine how to conserve this water.

Procedure

1 Obtain permission from an adult to do this activity at home.

2 **Measure** If you usually take a shower, the next time you take a shower, keep track of how long you were in the shower. Record this time.

_____

3 **Measure** Find out how much water you used for your shower. Collect water from the shower for one minute in a bucket. Measure this volume and multiply it by the number of minutes you showered. Record this volume.

_____

4 **Measure** If you take a bath instead of a shower, measure how much water is in the bathtub.

_____

5 **Apply** How much water will you use showering or bathing for a week? For a month?

_____

_____

6 How could you use less water to shower or bathe?

_____

_____

Dirty Air

1 Using a plastic knife, smear a thin layer of petroleum jelly on an index card.

2 Holding the edges of the index card, carefully place the card in a corner of the room.

Materials

- petroleum jelly
- index card
- plastic knife

3 **Observe** What does the index card look like after one day? After one week?

_____

_____

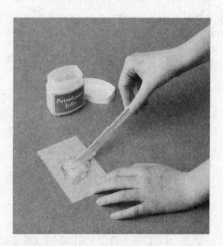

4 **Infer** How does the petroleum jelly help you track air pollution?

_____

_____

_____

_____

5 **Form a Hypothesis** Would you expect more air pollution near a road or away from a road? Why?

_____

_____

_____

Name _____ Date _____

How does the angle of sunlight affect temperature?

Form a Hypothesis

What happens to the temperature of Earth when sunlight reaches it at different angles? Write your answer as a hypothesis in the form "If the angle of the sunlight increases, then . . ."

_____

_____

Materials

- scissors
- 3 sheets of black construction paper
- masking tape
- 3 pieces of cardboard
- 3 thermometers
- protractor

Test Your Hypothesis

1 ⚠ **Be Careful.** Cut a slot for the thermometers in the middle of each piece of construction paper.

2 Tape each sheet of construction paper to one of the pieces of cardboard.

3 Place a thermometer into each slot so the bulb is between the construction paper and the cardboard.

4 Tape the thermometers in place. Place the thermometers in the shade until they read the same temperature. Record this temperature.

Step 3

5 ⚠ **Be Careful.** Do not look directly at the Sun. Put the thermometers in the sunlight as shown.

6 **Record Data** Every two minutes, record the temperature shown on each thermometer.

Step 5

Temperature Readings		
Thermometer 1	**Thermometer 2**	**Thermometer 3**

Draw Conclusions

7 What are the independent and dependent variables in this experiment?

_____

_____

8 **Interpret Data** On a separate sheet of paper, graph the change in temperature over time for each thermometer setup. Which thermometer's temperature rose fastest?

_____

Explore More

You know that sunlight warms Earth's surface. Which is warmed faster by sunlight—soil or water? Form a hypothesis, design an experiment to test it, record your data, and communicate your results.

_____

_____

_____

_____

Open Inquiry

What happens to the temperature of soil and water overnight? Think of your own question about how land and water temperature will change when the Sun goes down. Make a plan and carry out an experiment to answer your question.

My question is:

_____

_____

How can I test it?

_____

_____

_____

_____

My results are:

_____

_____

_____

_____

How do different colors affect temperature?

Make a Prediction

Make a prediction about how temperature will be affected by the color of construction paper covering a thermometer.

_____

_____

Materials

- thermometer
- 3 colors of construction paper
- tape

Test Your Prediction

1 **Measure** Tape a piece of construction paper loosely around a thermometer. Place the thermometer in a sunny window for five minutes. After five minutes, what is the temperature?

_____

_____

2 **Measure** Repeat step 1 using 2 different colors of construction paper. What temperatures were measured?

_____

_____

Draw Conclusions

3 **Communicate** Did your results support your prediction?

_____

_____

_____

_____

Name _____ Date _____

Air Pressure and Volume

Materials
- small plastic bag
- plastic container
- pencil

① **Make a Model** Set up the bag and container as shown. Make sure your setup is sealed.

② **Observe** Have a partner place both hands on the container and hold it firmly. Slowly push the bag into the container. Did the volume or the amount of air change as you pushed down? How does it feel? Why?

_____

_____

_____

_____

③ Pull the bag back out of the container. Using a pencil, carefully poke a hole in the plastic bag.

④ **Observe** Push the bag into the container again while holding your hand near the hole in the bag. Did the volume or the amount of air change as you pushed down? How did it feel? Why?

_____

_____

_____

_____

Communicate

When scientists complete an experiment, they **communicate** their results. When you **communicate**, you share information with others. You may do this by speaking, writing, drawing, singing, or dancing.

▶ **Learn It**

In the following activity, you will test whether air can lift a notebook off the table. Keep notes that include your hypothesis, materials, observations, and conclusion.

Scientists often try new experiments based on work that other scientists have done. If you **communicate** the details of your experiment, other students can do experiments based on yours. Writing down exactly what you did also lets you plan more experiments with different materials and different variables. If you get an unexpected result or disprove your hypothesis, you should **communicate** that as well.

▶ **Try It**

1 You know that air has weight and takes up space. Do you think air in a balloon will be able to lift a notebook off a table? If it can, how high will the notebook rise?

_____

_____

2 Tape two balloons to a notebook so the ends of the balloons stick out. Flip the notebook over so it is on top of the balloons.

Name _____ Date _____

3 Blow into one of the balloons. What happens to the notebook? Fill both balloons with as much air as you can.

_____

_____

Materials
• notebook
• balloons
• tape
• ruler

4 Using a ruler, measure the height between the table and notebook. Record the results.

_____

_____

_____

5 **Communicate** Exchange data about the height to which air was able to raise your notebook.

6 Using the data from your classmates, figure out the average height that your class was able to lift the notebooks. Make a chart to compare your results.

7 **Communicate** Who was able to raise their notebooks the highest? Was anyone unable to lift it? Discuss any problems that occurred or improvements that could be made to lift the notebook higher.

_____

_____

_____

_____

_____

▶ **Apply It**

Think about how you could use air to lift the book even higher. What would happen if you used bigger balloons? If you placed smaller balloons under each corner of the notebook? How heavy of a book could you lift using these materials?

_____

_____

_____

_____

Plan a new experiment. Test your hypothesis and draw conclusions about using the power of air to lift objects. Then communicate to the class the results of your experiment by writing a report, drawing a cartoon strip, or composing and singing a song!

_____

_____

_____

_____

Name _____ Date _____

How much rain falls in your community?

Purpose
To measure the amount of rainfall in your community.

Procedure

Materials

- scissors
- carton
- masking tape
- baking pan
- ruler

1. ⚠ **Be Careful.** Use the scissors to cut the top off the carton.

2. Using tape, attach the carton to the baking pan. Then put the pan on the ground outside in an open area.

3. **Measure** Check the carton at the same time every day. If there is water in it, measure the height of the water in centimeters.

Step 2

4. **Record Data** Write down the daily results on a table. Then empty the carton and put it back in the same spot outside.

_____

_____

Step 4

Day 1	Day 2	Day 3	Day 4	Day 5

© Macmillan/McGraw-Hill

Draw Conclusions

5 **Interpret Data** Design and complete a graph to display your results.

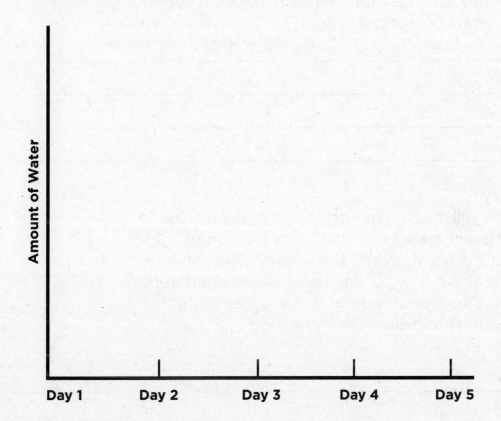

Amount of Water

Day 1   Day 2   Day 3   Day 4   Day 5

6 **Use Numbers** Calculate the volume of water you measured by multiplying the rainfall by the area of the container. Convert your total measurement of cubic centimeters of precipitation into liters.

_____

_____

Name _____ Date _____

Explore More

How close were your results to an official rain measurement for your area? Were there any problems that you ran into with the experiment? How could you improve your data collection?

_____

_____

_____

_____

Open Inquiry

How accurate will measurements be if rain gauges are placed in different areas? Can the things the rain gauge is near affect the accuracy of calculations? Think of your own question about how putting rain gauges in different positions can affect your findings. Make a plan and carry out an experiment to answer your question.

My question is:

_____

_____

How I can test it?

_____

_____

_____

My results are:

_____

_____

_____

_____

Do different containers affect precipitation measurements?

Materials
• ruler
• 3 different containers
• graph paper

Form a Hypothesis

Form a hypothesis on whether the type of container used as a rain gauge will affect the amount of precipitation measured.

_____

_____

Test Your Hypothesis

1 **Experiment** Use three containers that are different sizes and shapes as rain gauges. Make sure to place the containers in similar locations.

2 **Measure** Use the ruler to measure the amount of precipitation each day at the same time. Record the data in a data table. How much precipitation did you measure in each container?

_____

_____

_____

3 **Interpret Data** Make a bar graph to compare the total amount of precipitation measured during the two week period in each of the containers.

Draw Conclusions

4 **Communicate** Did the results of the experiment support your hypothesis?

_____

_____

Precipitation

Containers

© Macmillan/McGraw-Hill

Name _____ Date _____

Types of Clouds

1 **Observe** Look for clouds in the sky. How many different types of clouds do you see?

_____

_____

2 **Classify** Do the clouds that you see look like cirrus, cumulus, or stratus clouds?

_____

_____

3 Continue your data collection for one week.

Cloud Observation			
	Cirrus	**Cumulus**	**Stratus**
Day 1			
Day 2			
Day 3			
Day 4			
Day 5			
Day 6			
Day 7			

4 Which type of cloud did you see most frequently?

_____

5 Write a report about the types of clouds that you saw. Do you think you would get different results at a different time of year? Explain.

Name _____ Date _____

Structured Inquiry

How can you tell that water vapor is in the air?

Form a Hypothesis

Cobalt chloride paper is blue in dry air and turns pink in air that has water vapor in it. Write a hypothesis in the form "If water evaporates, then cobalt chloride paper near or above the water will . . ."

_____

_____

Materials

- 2 bottles
- scissors
- tape
- cobalt chloride paper
- 2 plastic cups
- sheet of paper

Test Your Hypothesis

1. ⚠ **Be Careful.** Cut the tops off the 2 bottles.

Step 1

2. Tape 1 strip of cobalt chloride paper in the bottom of each bottle.

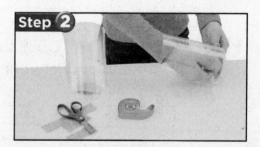
Step 2

3. Place 1 bottle upside down over an empty plastic cup. Fill the other plastic cup half-full of water. Place a bottle upside down over that cup.

Step 3

© Macmillan/McGraw-Hill

Name _____ Date _____

④ Tape a third strip of cobalt chloride paper to a sheet of paper. Leave it in open air.

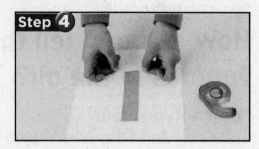

Step ④

⑤ **Observe** Examine the color of the strips of cobalt chloride paper.

⑥ **Record Data** Write down any changes in the color of the cobalt chloride paper.

_____

_____

Draw Conclusions

① **Use Variables** Identify the variables in this experiment. What purpose does the cobalt chloride paper that is taped to the piece of paper serve?

_____

_____

_____

_____

② **Draw Conclusions** Does the evidence from your observations support your hypothesis?

_____

_____

Be a Scientist

Guided Inquiry

Does surface area affect rate of evaporation?

Form a Hypothesis

You have already learned water vapor can be detected in the air. Does water evaporate faster from a body of water with a bigger surface area? Write your answer as a hypothesis in the form "If you increase the surface area of water, then the evaporation rate will . . ."

_____

_____

Test Your Hypothesis

Design a plan to test your hypothesis. Write out the materials, resources, and steps that you need to take. Record your results and observations as you follow your plan.

Draw Conclusions

Did your test support your hypothesis? Why or why not? Present your results to your classmates.

_____

_____

© Macmillan/McGraw-Hill

Name _____ Date _____

Open Inquiry

What effect does wind have on the evaporation rate of water? Come up with a question to investigate. Design an experiment to answer your question. Your experiment must be organized to test only one variable or one item being changed. Your experiment must be written so that another can complete it by following your instructions.

My hypothesis is:

_____

How can I test it?

_____

_____

_____

_____

_____

My conclusions are:

_____

_____

_____

_____

What happens when air masses of different temperatures meet?

Form a Hypothesis

What happens to an air mass when it meets another air mass of the same temperature or of a cooler temperature? Write your answer as a hypothesis in the form "If an air mass meets another air mass of the same or of a cooler temperature, then . . ." Like air, water flows and carries heat. Using water as a model for air can help you test your hypothesis.

_____

_____

_____

_____

_____

Materials

- scissors
- cardboard
- clear plastic box
- aluminum foil
- cold water
- 2 containers
- warm water
- food coloring

Test Your Hypothesis

1 ⚠ **Be Careful.** Using scissors, cut the cardboard so it fits tightly in the clear box. Wrap the cardboard in aluminum foil.

2 Pour 4 cups of cool water into one container and 4 cups of warm water into the other one. Place a few drops of blue food coloring into the cool water and red into the warm water.

3 Hold the cardboard tightly against the bottom of the box. Pour the cool water on one side and the warm water on the other.

© Macmillan/McGraw-Hill

Name _____ Date _____

4. **Observe** Watch the box from the side as you remove the cardboard. Describe what happens.

Step 4

_____

_____

5. Now repeat the same test with warm water in both containers and food coloring in only one.

_____

_____

Draw Conclusions

6. What are the variables in this experiment?

_____

_____

7. **Infer** Which test looked like it formed storms? Why?

_____

_____

_____

Explore More

Will a greater difference in temperature between the warm and cold water increase the observable effects? Form a hypothesis and test it.

_____

_____

_____

_____

Open Inquiry

What would happen with different amounts of water?
Think of your own question about how the amounts of
water affect what happens. Make a plan and carry out an
experiment to answer your question.

My question is:

_____

_____

How I can test it:

_____

_____

_____

_____

My results are:

_____

_____

_____

_____

Name _____ Date _____

How does mixing rate affect the interaction of water masses with different temperatures?

Materials

- scissors
- warm and cool water
- cardboard
- aluminum foil
- 2 containers
- food coloring
- clear plastic box

Form a Hypothesis

Form a hypothesis about whether the mixing rate will affect how two water masses with different temperatures will interact.

_____

_____

_____

Test Your Hypothesis

1 Measure Cut the cardboard so it fits tightly in the clear box. Wrap the cardboard in aluminum foil.

2 Observe Pour 4 cups of cool water into one container and 4 cups of warm water into the other one. Put a few drops of food coloring into the cool water container. Hold the cardboard tightly against the bottom of the box. Pour the cool water into one side and the warm water into the other side. Watch the box from the side as you remove the cardboard very slowly. What do you observe?

_____

_____

3 Observe Repeat step 2, but this time remove the cardboard divider very quickly. What do you observe?

_____

_____

Tornado in a Bottle

Materials

- two 2-liter plastic bottles
- duct tape or a connector
- water
- paper towels

1. Fill a 2-liter plastic bottle one-third full of water.

2. Place an empty 2-liter plastic bottle upside down over the mouth of the first bottle. Use masking tape or a connector to join the two bottles together.

3. **Make a Model** Holding the bottles by the necks, flip them upside down so the bottle with the water in it is now on top. Place the bottles on a desk.

4. **Observe** What do you see?

   _____

   _____

5. How is this model similar to the movement of wind in a tornado?

   _____

   _____

   _____

Name _____ Date _____

How does distance from an ocean affect temperature?

Make a Prediction

San Francisco, CA is closer to the Pacific Ocean than Stockton, CA. Make a prediction about how distance from an ocean affects the temperature of a city.

_____

_____

_____

Test Your Prediction

1. Use the temperature data in the charts to compare the monthly high temperatures of the two cities.

_____

_____

_____

2. Use the temperature data in the charts to compare the monthly low temperatures of the two cities.

_____

_____

Average High Temperature (°F)		
	San Francisco	Stockton
Jan.	55.7	53.4
Feb.	59.1	60.5
Mar.	61.3	65.9
Apr.	63.9	72.9
May	66.8	81.0
Jun.	70.0	88.4
Jul.	71.4	94.1
Aug.	72.1	92.5
Sep.	73.5	88.2
Oct.	70.2	78.4
Nov.	62.9	64.2
Dec.	56.4	53.7

Average Low Temperature (°F)		
	San Francisco	Stockton
Jan.	42.4	37.7
Feb.	44.9	40.5
Mar.	46.1	42.6
Apr.	47.6	46.1
May	50.1	51.6
Jun.	52.6	57.0
Jul.	53.9	60.4
Aug.	54.9	59.8
Sep.	54.7	57.2
Oct.	51.8	50.2
Nov.	47.3	42.2
Dec.	43.1	37.5

© Macmillan/McGraw-Hill

Draw Conclusions

3 **Interpret Data** Which city has the greater change in temperature during the year? Which city has the smaller change in temperature during the year?

_____

_____

_____

4 **Infer**. How might the ocean affect the temperature changes in these cities?

_____

_____

5 **Communicate** Write a report explaining how the data for these two cities either support or do not support your prediction. Would examining data for more cities improve the accuracy of your prediction?

_____

Explore More

Write a prediction explaining how being near an ocean will affect another weather variable. Collect and compare weather data for both cities. Write a report explaining how the data support or do not support your prediction.

_____

_____

_____

Name _____ Date _____

Open Inquiry

How does elevation above sea level affect temperatures?
Think of your own question about how a city's elevation
affects its temperatures. Make a plan and carry out an
experiment to answer your question.

My question is:

_____

_____

How I can test it:

_____

_____

_____

_____

My results are:

_____

_____

_____

_____

Name _____ Date _____

Does the location of a mountain range affect rainfall?

Materials

- map of the United States
- Internet
- science reference books

Form a Hypothesis

Form a hypothesis about whether a mountain range will affect the average annual rainfall in a city.

_____

_____

Test Your Hypothesis

1 **Observe** Find two cities that are on opposite sides of a large mountain range. Use the Internet or other science reference books to find the average annual rainfall of the two cities.

2 **Use Numbers** What is the average annual rainfall in the city on the west side of the mountains?

_____

3 **Use Numbers** What is the average annual rainfall in the city on the east side of the mountains?

_____

4 **Interpret Data** Which city had the lower average annual rainfall?

_____

_____

Draw Conclusions

5 **Infer** Why do you think one city had a lower average annual rainfall?

_____

_____

© Macmillan/McGraw-Hill

Name _____ Date _____

Climate and Rain Shadow

1. **Make a Model** Honolulu and Kailua are on opposite sides of a mountain range in Hawaii. In order to figure out the location of the cities compared to the mountain range, what weather information would you need?

_____

_____

Cities	Annual Temperature (°F)	Annual Precipitation (Inches)
Honolulu	77.1	20.73
Kailua	70.2	118.97

2. Which location receives more rain?

_____

3. Which location is warmer?

_____

4. **Infer** Which city is on the windward side of the mountain range?

_____

5. **Infer** Which city is in the rain shadow?

_____

What keeps Earth moving around the Sun?

Form a Hypothesis

If you let go of a ball being swung in a circle, in what direction will the ball travel? Write a hypothesis in the form "If I let go of a ball being swung in a circle at a particular point, then . . . "

_____

_____

_____

_____

Materials

• tennis ball

• fabric

• string

• graph paper

Test Your Hypothesis

1. Place the tennis ball on the fabric and bring the four corners of the fabric together so they cover the ball. Then tie string around the four corners to form a pouch.

Step 1

2. ⚠ **Be Careful.** While holding the other end of the string, lean forward and slowly spin the ball in a circle near your feet.

Step 2

3 **Observe** Let go of the string. Watch the path that the ball takes.

4 **Record Data** Draw a diagram to show the path the ball took when you let it go.

[blank box]

5 Repeat the experiment, letting go of the ball at three different spots on the circle. Where does the ball go?

_____

_____

Draw Conclusions

6 Did the experiment support your hypothesis? Why or why not?

_____

_____

7 If this activity models the movement of Earth around the Sun, what do you, the ball, and the string represent?

_____

_____

Explore More

What results would you expect if you repeated the experiment using a lighter ball? Form a hypothesis, do the experiment, record your data, and write a report.

_____

_____

_____

_____

Open Inquiry

What is the relationship between weight and distance traveled when you release the ball in this experiment? Is there a relationship you can find? Think of your own question about how weight affects distance traveled. Make a plan and carry out an experiment to answer your question.

My question is:

_____

_____

How I can test it:

_____

_____

_____

My results are:

_____

_____

_____

© Macmillan/McGraw-Hill

Name _____ Date _____

What do we learn from shadows?

Materials
- sheet of paper
- ruler
- clay
- transparent tape
- string
- protractor

Purpose

Find out how shadows change during the day.

⚠ **Be Careful.** Do not look directly at the Sun at any time.

Procedure

1 Draw two lines on a sheet of paper (top to bottom and left to right) to make four quarters. Label map directions on the paper: N at the top center, S at the bottom center, W at the center of the left side, and E at the center of the right side.

2 Stand a pencil upright in a blob of clay. Place the clay in the center of your paper.

3 Place the paper on a flat surface where the Sun will shine on it all day. Place it so the N points north.

4 **Observe** At 10 A.M. carefully trace the pencil's shadow. Put a heavy dot on the tip of the shadow tracing. Repeat at 11 A.M., 12 noon, 2 P.M., and 3 P.M., and record your work.

Draw Conclusions

5 When was the shadow longest? Shortest?

_____

_____

6 **Interpret Data** What is the relationship between the shadow and the Sun's location?

_____

_____

Name _____ Date _____

Seasons and Earth's Tilt

Materials

Materials
- modeling clay
- toothpick
- flashlight
- pencil

1. Using modeling clay, make a sphere to represent Earth. Then make a base for the sphere.

2. ⚠ **Be Careful.** Push a toothpick through the sphere to represent Earth's axis. Use a pencil to draw a line around the center to represent the equator.

3. Hold the sphere so the toothpick in it is straight up and down, then tilt the sphere so the top of the toothpick is at an angle of about 23° and push the bottom of the toothpick into the base.

4. **Observe** Aim a flashlight at the sphere so the end of the toothpick points away from you. Describe how the light spreads over the sphere. What would the seasons be in the Northern and Southern Hemispheres?

_____

_____

_____

5. **Observe** Now shine the flashlight so the end of the toothpick points towards you. Describe how the beam of light spreads over the sphere. What would the seasons be in the Northern and Southern Hemispheres?

_____

_____

_____

Name _____ Date _____

Use Numbers

When scientists **use numbers**, they add, subtract, multiply, divide, count, or put numbers in order to explain and analyze data.

The orbits of each planet in the solar system have different radii. This means each planet takes a different amount of time to revolve around the Sun. As the radius of the planet's orbit increases, the revolution time increases. What would your age be if you lived on a different planet?

▶ **Learn It**

The diagram of the planets shows the time each planet takes to revolve around the Sun in Earth days or years. Scientists **use numbers** to compare the revolution time of the other planets in our solar system to Earth. You can do that by dividing the revolution time of a planet by the revolution time of Earth.

For example, it takes Earth $365\frac{1}{4}$ days to travel around the Sun. Mars takes 687 days to complete its revolution. If you divide the length of time it takes Mars to make a revolution by the length of time it takes Earth to make a revolution, you get 1.88. Mars takes almost twice as long as Earth to complete one revolution.

If you were 62 years old in Earth years, how old would you be in Mars years? The ratio of Mars's revolution to Earth's is 1.88. Divide your age by the Earth-planet ratio to calculate your age on a specific planet.

Rotation Data

Planet	Revolution (days)	Earth–Planet Ratio	Age on Planet
Mercury			
Venus			
Earth	365	1	62
Mars	687	1.88	33
Jupiter			
Saturn			
Uranus			
Neptune			

Name _____ Date _____

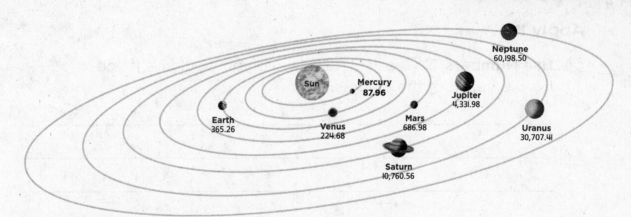

▶ **Try It**

❶ Record the revolution data from the diagram into the chart below.

Planet	Revolution (days)	Earth-Planet Ratio	Age on Planet
Mercury			
Venus			
Earth			
Mars			
Jupiter			
Saturn			
Uranus			
Neptune			

❷ **Use Numbers** Calculate the Earth-planet ratio for all of the planets.

❸ If you are 6 years old in Earth years, how old would you be in Mars years?

_____

▶ **Apply It**

④ **Use Numbers** Now calculate how old you would be if you lived on each of the planets.

_____

_____

_____

⑤ On which planet would you be the oldest in that planet's years? On which planet would you be the youngest?

_____

⑥ What can you infer about the revolution time of the planet and the age you would be on that planet?

_____

_____

What makes the Moon appear to change shape?

Materials

- lamp
- large ball

Purpose

To model changes in the appearance of the Moon as seen from Earth.

Procedure

1 **Make a Model** You represent an observer on Earth. A classmate uses a lamp to represent the Sun. A classmate with a ball represents the Moon.

2 Face the classmate with the lamp. Have your other classmate hold the ball directly between you and the lamp.

3 **Observe** Have your classmate turn the lamp on. How much of the surface of the ball is lit? Record what you see.

Name _____ Date _____

④ **Observe** Have your classmate with the ball move one eighth of the way around you. Turn to face the ball and record what you see.

_____

_____

_____

⑤ Repeat step 4 until your classmate returns to the starting position. Record your observations.

_____

_____

_____

_____

Draw Conclusions

⑥ What causes the changes in Moon's appearance as seen from Earth?

_____

_____

_____

_____

⑦ What happened to the shape of the ball that represented the Moon during this experiment?

_____

_____

_____

Explore More

Make a mark on one side of the ball. Repeat the procedure keeping the marked side towards the light. How much sunlight do the marked and unmarked sides receive?

_____

_____

Open Inquiry

What happens to the Moon during the daytime? Think of your own question about what happens to the Moon during the daytime. Make a plan and carry out an experiment to answer your question.

My question is:

_____

_____

How I can test it:

_____

_____

_____

_____

_____

My results are:

_____

_____

_____

_____

Name _____ Date _____

How does the Moon look from Earth?

Materials
- tennis ball
- flashlight

Purpose

The changing appearance of the Moon is based on the positions of Earth, the Moon, and the Sun. How do you think you can position a ball, representing the Moon, and a flashlight, representing the Sun, to create a full moon and a half moon?

_____

_____

_____

Procedure

1. **Observe** Turn on the flashlight and shine it at the tennis ball in a dark location. If the tennis ball represents the Moon, the flashlight represents the Sun, and you represent the Earth, how should you position the ball and flashlight so that you, as Earth, would view a full Moon?

_____

_____

2. **Observe** Find the positions of the Sun and Moon for a half Moon. What position do the ball and flashlight have to be in for you to view a half-moon?

_____

3. **Infer** Where does all of the light that is seen from the Moon come from?

_____

Name _____ Date _____

Eclipses

Materials

- large ball
- tennis ball
- lamp

1 **Make a Model** You will represent Earth using a large ball. One classmate will use a lamp to model the Sun. Another classmate will use a tennis ball to represent the Moon.

2 Begin by facing the classmate with the lamp. Rotate the Moon through one month of movement in its orbit.

3 What positions of the Moon, Earth, and Sun produce eclipses?

_____

_____

_____

_____

Name _____ Date _____

How far apart are the planets?

Materials

- paper towels
- markers
- ruler
- tape

Purpose

To make a model that shows the distances between the planets using Astronomical Units (AU), where one AU equals the average distance between Earth and the Sun. This distance is about 149,591,000 kilometers (92,960,000 miles).

Procedure

1. Let the length of each paper towel equal 1 Astronomical Unit. Using the chart, lay out the number of paper towels you need to show the distance from the Sun to Neptune.

Distance of the Planets from the Sun	
Planet	**Distance in AU**
Mercury	.39
Venus	.7
Earth	1
Mars	1.5
Jupiter	5.2
Saturn	9.5
Uranus	19.2
Neptune	30

2 **Make a Model** Mark the location of the Sun at one end. Then measure the distance that each planet would be from the Sun and draw the planet on the paper towel.

Step **2**

Draw Conclusions

3 **Interpret Data** Compare the distances between Mercury and Mars, Mars and Jupiter, and Jupiter and Neptune. Which are farthest apart?

_____

_____

4 **Infer** What can you conclude about the distances between the planets in the solar system?

_____

_____

_____

Explore More

Your model has all of the planets in a line. How could you make a model to show the positions of the planets at a specific time? Write instructions that others can follow to make the model.

_____

_____

_____

Open Inquiry

In the models and pictures you have seen of the solar system, what were the sizes of the planets on those scales? Think of your own question about the size of the planets in your paper-towel model. Make a plan and carry out an experiment to answer your question.

My question is:

_____

_____

How I can test it:

_____

_____

_____

My results are:

_____

_____

© Macmillan/McGraw-Hill

How do the masses of objects in the solar system compare?

Materials

• calculator

Purpose

Find out about the masses of the objects that make up the solar system, which include the Sun, the planets that orbit the Sun, moons, asteroids, and other objects.

Procedure

1 Use the Internet to find out the approximate masses of the objects that make up the solar system. List the masses in the table below. Be sure to use the same units for all masses.

Object	Mass
Sun	
Mercury	
Venus	
Earth	
Mars	
Jupiter	
Saturn	
Uranus	
Neptune	
moons	
asteroids	
other objects	

2 **Use Numbers** Calculate the total mass of the solar system.

_____

2 **Draw Conclusions** What did you learn about the relative masses of different objects in the solar system?

_____

_____

_____

Name _____ Date _____

Planet Sizes

1 **Use Numbers** Using a scale of 2,000 kilometers = 1 centimeter, find the diameter of each of the planets in centimeters.

Materials
- poster board
- ruler
- scissors

	Diameter in kilometers	Model size in centimeters
Mercury		
Venus		
Earth		
Mars		
Jupiter		
Saturn		
Uranus		
Neptune		

2 **Make a Model** Using a ruler and scissors, cut circles out of poster board to show the sizes of the planets. Then label each planet.

3 Arrange the planets in order from nearest to farthest from the Sun.

4 How do the sizes of the inner and outer planets compare?

_____

_____

_____

How does distance affect how bright a star appears?

Form a Hypothesis

If one star gives off more light than another star, but they appear to be the same brightness to an observer, what does this mean about the distance of the stars from the observer? Write your answer as a hypothesis in the form "If one star gives off more light than another star but both appear to be the same brightness to an observer, then . . ."

_____

_____

_____

_____

_____

Materials

- 2 flashlights
- tissue paper
- rubber band
- masking tape
- meterstick

Procedure

1 Cover the front of one flashlight with a few layers of tissue paper. Place a rubber band around the paper.

2 **Make a Model** Let each flashlight represent a star. Place a strip of masking tape on the floor. Have two classmates stand behind the tape and turn the flashlights on.

© Macmillan/McGraw-Hill

3 Select the flashlight that appears brighter and have the classmate holding it slowly move away from you. When do the two stars appear to have the same brightness? Measure the distance.

Distance of flashlight 1 _____

Distance of flashlight 2 _____

Draw Conclusions

4 **Infer** What factors affect how bright a star looks to an Earth observer?

_____

_____

5 **Communicate** Did your classmates see the stars as having the same brightness at different distances? What might that mean about individual observations of stars?

_____

_____

Explore More

Two stars give off the same amount of light, but one
star looks dimmer. Form a hypothesis and use models to
test your prediction. Collect data and communicate your
results.

_____

_____

_____

_____

_____

Open Inquiry

One thing that affects a star's appearance in the sky is the
star's nearness to Earth. Think of your own question about
how some stars look from here on Earth. Make a plan and
carry out an experiment to answer your question.

My question is:

_____

_____

How I can test it:

_____

_____

_____

My results are:

_____

_____

Name _____ Date _____

What causes a star's color?

Materials

• pencil

Make a Prediction

Stars are found in many different colors, temperatures, sizes, and degrees of brightness. Some characteristics of a star, such as brightness, are related to the star's distance from Earth. A star's color, however, can be linked to its temperature. How can you use the table below to estimate the temperature of a star by observing its color?

_____

_____

Temperature (Kelvin)	Color
>27000	blue-white
27000-11000	blue-white
11000-7200	white
7200-6000	yellow-white
6000-5100	yellow
5100-3700	orange
<3700	red

Test Your Prediction

1. **Observe** Look at the chart that illustrates the various colors and temperatures (in Kelvins) of a star. What temperature range would you predict a star that appears orange would be?

_____

2. What do you think is the temperature range of the Sun?

_____

Name _____ Date _____

Expanding Universe

Materials
- balloon
- binder clip
- stickers

1 **Make a Model** Blow the balloon up a little. The balloon represents the universe shortly after the big bang. Place stickers on it to represent galaxies.

2 **Observe** Blow the balloon half-full of air. What happens to the size of the stickers? To the distance between the stickers?

_____

_____

_____

3 **Observe** Blow the balloon full of air. What happens to the size of the stickers? To the distance between the stickers?

_____

_____

_____

Name _____ Date _____

Structured Inquiry

How do craters form?

Form a Hypothesis

You know that craters form when an object in space hits another object. Does the size of an object affect the size of the crater it forms? Write your answer as a hypothesis in the form "If a larger object hits, then . . ."

_____

_____

Test Your Hypothesis

1. Cover the floor with newspaper and place a pan on the paper.

2. **Make a Model** Fill the pan with cocoa powder to about 1 cm. Gently tap the pan until the cocoa powder is smooth. Using the spoon, shake white flour on top to represent topsoil.

Step 2

③ By wrapping a cut rubber band around the marble, measure the diameter of three marbles of different sizes.

	Diameter
Marble 1	
Marble 2	
Marble 3	

④ Drop the largest marble from 20 cm above the pan. Measure the diameter of the crater and record your data.

Step ④

⑤ Repeat step 4 for the other 2 marbles. Make sure each marble falls in a different area of the pan.

	Diameter
Crater 1	
Crater 2	
Crater 3	

Draw Conclusions

⑥ **Analyze Data** How does the diameter of the crater compare to the diameter of the marble?

_____

_____

_____

_____

7 What did you see at the crater sites? Why did this happen?

_____

_____

_____

_____

8 How is this model similar to what happens when an object hits the surface of the Moon?

_____

_____

_____

_____

9 What are the controlled, independent, and dependent variables?

_____

_____

_____

_____

Guided Inquiry

How does height affect crater size?

Form a Hypothesis

You now know the effect that objects of different sizes have on crater formation. What happens when similar sized objects hit from different heights? Write your answer as a hypothesis in the form "If an object hits from a greater height, then . . ."

_____

_____

Test Your Hypothesis

Design an experiment to test your hypothesis. Write out the materials you need and the steps you will take. Record your results and observations.

_____

_____

_____

Draw Conclusions

What were your independent and dependent variables? Did your experiment support your hypothesis?

_____

_____

Open Inquiry

What effect does the surface material have on crater formation? Think of a hypothesis and design an experiment to answer it. Your experiment must be organized to test only one variable. Keep careful notes as you do your experiment so another group could repeat the experiment by following your instructions.

My hypothesis is:

_____

_____

How I can test it:

_____

_____

_____

My conclusions are:

_____

_____

© Macmillan/McGraw-Hill

Name _____ Date _____

Which has more matter?

Form a Hypothesis

Which do you think has more matter—the balloon or the tennis ball? Does having more matter make an object larger? Does it make it heavier? Write your answer as an hypothesis in the form: "If matter increases then an object will . . ."

_____

Test Your Hypothesis

Materials

- inflated balloon
- tennis ball
- container of water
- tape
- equal pan balance

1. **Measure** Place one object underwater. Record how high the water rises with some tape, then remove the object. Do not spill any water! Next place the second object in the water and record the water level—this is your dependent variable.

   Water Level

   With the balloon _____

   With the tennis ball _____

   Balance Measurements

   Balloon _____

   Tennis ball _____

Step 1

2. Place the objects on either side of the equal pan balance. Which is heavier? This is your independent variable.

   _____

3. Repeat all your measurements to verify your answers.

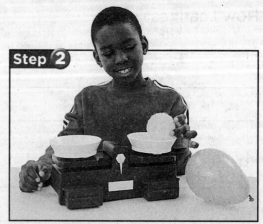

Step 2

© Macmillan/McGraw-Hill

Draw Conclusions

4 **Interpret Data** Did the heavier object also raise the water higher? Why or why not? Which object has more matter? Did the test support your hypothesis?

_____

_____

_____

_____

Explore More

What if you were given a large bag of popped popcorn and a small bag with an equal number of popcorn kernels? Which do you think would have more matter? Form a hypothesis and test it. Then analyze and write a report of your results.

_____

_____

_____

_____

Name _____ Date _____

Open Inquiry

Design an experiment to compare the mass of two objects.

My question is:

_____

_____

How I can test it:

_____

_____

My results are:

_____

_____

Does the same volume mean the same mass?

Form a Hypothesis

Do you think that two objects of the same size will have the same mass? Form a hypothesis to answer that question.

_____

_____

Test Your Hypothesis

1 Work with a small group of students. Get two identical balloons from your teacher.

2 **Experiment** Add water to one of the balloons and tie it off. Add air to the other balloon until it is the same size as the first balloon. Tie it off too.

3 **Measure** Place the balloon containing water on one side of the balance. Place the balloon containing air on the other side. In the space below, show what you observe.

Materials

- 2 identical balloons
- water
- balance

4 **Interpret Data** Which balloon contains more matter? How do you know?

_____

_____

Name _____ Date _____

Too Dense to Float?

- Piece of paper towel
- 1 plastic cup
- 1 foam cup

1 **Make a Model** You will need three models. The first should be a triangle cut from a paper towel. The second will be a plastic cup. The third will be a foam cup.

2 Place all three objects in water. The triangle should lie flat on the water. The two cups should be placed with the open end down. Which ones sink? Record your answers.

_____

3 Add liquid detergent to the water to break up the surface tension. Do any models sink?

_____

4 Have your teacher poke evenly spaced holes in each model. Do any models sink now?

_____

_____

_____

5 **Infer** Which models could float simply because of their density? Which models needed help from their shape or were on top of the water due to surface tension?

_____

_____

_____

© Macmillan/McGraw-Hill

Infer

When scientists see something interesting they make careful records of what they find. They then think about what they have seen and try to **infer** what it means about the world as a whole. Understanding why something happens allows you to draw conclusions about how objects behave or what properties they may have.

Think about a material that is denser than water. It has more mass per milliliter than water—there is more matter in the same amount of space. Will a material denser than water float? By watching what happens to such objects made of such materials, you can infer what keeps objects afloat in general.

▶ **Learn It**

When you **infer**, you form an opinion after analyzing recorded data. It's easier to analyze data if you organize the information on a chart or in a graph. That way you can quickly see differences among data and make inferences. Most metal objects, such as a spoon or nail, sink quickly. This is because they are denser than water. There are large metal boats, however, that regularly carry heavy cargo across the ocean. How can they stay afloat? We will make several model boats to help us infer an answer. The boats will be made out a material denser than water—the metal aluminum.

Name _____ Date _____

▶ **Try It**

1 Take a sheet of aluminum foil. Use the foil to make a boat. Experiment with different designs. Draw a picture of the boat in a chart.

Materials
- aluminum foil
- paper clips
- tank of water

	Picture	Number of Clips	How did it work?
Boat 1			
Boat 2			
Boat 3			
Boat 4			

2 Float the boat in a large pan of water. Place large paper clips into the boat and record what happens. How many paper clips can the boat hold before it completely sinks? Try to **infer** why the boat is sinking.

© Macmillan/McGraw-Hill

▶ **Apply It**

1 Record the data and results from two other students in your chart.

2 Now it is time to analyze your data. Do you notice any pattern between the design of the boat and the number of paper clips?

_____

_____

3 As a class, design a boat that would carry the most paper clips possible. Use a final piece of aluminum foil to make the boat and record how many paper clips it can hold. Did this boat hold more paper clips than the others?

_____

4 Think about all the models you have seen. Did the ones that held more paper clips have anything in common? What was happening as more paper clips were added to the boat? Use your observations to **infer** what makes an object float. Communicate your opinions by writing down your conclusions.

_____

_____

_____

_____

_____

Name _____ Date _____

How can you know what's "inside" matter?

Purpose
You will examine four sealed boxes to determine what is inside them.

Procedure

1 **Observe** Examine the four boxes, but do not open them. You can lift them, shake them, listen to the noises they make, feel the way they shift when you move them, and so on. Don't forget to use the magnet and equal pan balance to learn more about what is inside the boxes. Record your observations.

_____
_____
_____
_____
_____
_____
_____
_____
_____
_____
_____
_____
_____

Materials

- 4 sealed, opaque boxes of different sizes, shapes, and colors.

- equal pan balance with set of masses

- magnet

Step **1**

Use with **Lesson 2**
Elements

Explore

2 **Infer** Try to determine
what is in each box.

Step **2**

_____

_____

_____

_____

_____

_____

_____

_____

Draw Conclusions

3 **Communicate** Describe what you think is in each box.

_____

_____

4 What evidence did you use to make your decisions?

_____

_____

_____

_____

_____

_____

5 When everyone in the class is finished, open the
boxes and reveal what is inside. Were your inferences
correct? Now that you know what is in each box,
explain any wrong guesses you made.

_____

_____

_____

_____

Explore More

What if you were the one to fill the boxes before this
experiment? What kind of items could you choose to
make the experiment easier? To make it harder? Choose
a few items that would fit inside the boxes. Now design a
series of tests that would prove that those items were the
ones inside the boxes.

Open Inquiry

Put an object in a box and seal it. Trade boxes with a
classmate. Design an experiment to find out what is inside
the box.

My question is:

_____

_____

How can I test it?

_____

_____

My results are:

_____

_____

© Macmillan/McGraw-Hill

What's in the box?

Materials
- shoebox
- 3 small items
- tape

Purpose
Infer the contents of a box by indirect methods to learn how objects too small to see can be studied.

Procedure

1 Work with a small group of students. Get a sealed shoebox from your teacher. The box contains three small items.

2 **Experiment** Perform several tests on the box to find out what is in it. Record the results in the table below.

What's in the Box?	
Test Used	**Observations**

3 **Interpret Data** From your observations, what three items do you think are in the box?

_____

_____

_____

4 **Infer** How do you think methods like this one help scientists learn about atoms?

_____

_____

_____

Name _____ Date _____

Inside Atoms and Molecules

1 **Make a Model** Use toothpicks to join eight large pink marshmallows (protons) to eight large green marshmallows (neutrons) to form the nucleus of an oxygen atom. Add eight small marshmallows around the outside as electrons.

2 Make another oxygen atom or share with another student. Use two pipe cleaners to join the atoms by two electrons. This is an oxygen molecule (O_2).

3 How do the shapes of your model atoms compare to the diagrams in this book?

_____

_____

_____

4 **Communicate** On another piece of paper, draw pictures of your atoms and molecule that show their actual shapes better.

5 In a molecule, electrons move and are sometimes traded between atoms. How could you represent this in your model?

_____

_____

_____

Name _____ Date _____

How can you tell if it is metal?

Materials

- safety goggles
- plastic, metal, and glass rods
- paper ties with steel wire
- wood toothpicks
- aluminum foil

Purpose

In this activity, you will observe, compare, and contrast metal and nonmetal objects. You will describe each object as strong or weak examples of several important properties.

Procedure

1 Prepare a table to record your observations. Label it as shown.

Property Used	Thermal Conductivity	Luster	Flexibility
plastic rods			
metal rods			
glass rods			
steel wire in paper ties			
wood toothpicks			
aluminum foil			

© Macmillan/McGraw-Hill

2 **Experiment** Test for thermal conductivity: Place each object half in the sun, or have under a lamp. Which materials seem hot to the touch on the non-lit half?

_____

3 Test for shiny luster: Look at the aluminum foil and sheet of paper. Which reflects more light?

_____

4 △ **Be Careful.** Wear goggles. Test for flexibility: Bend a tie with wire in half. Bend a toothpick in the same way. Which holds its new shape without breaking?

_____

Draw Conclusions

5 **Classify** Use your observations to decide if objects were strong or weak examples of the properties you tested.

_____

_____

_____

_____

_____

6 **Communicate** Based on your observations, summarize the properties of metals and nonmetals.

_____

_____

Explore More

Are the properties of all metals the same? Are some stronger examples of some properties, but not others? Plan and conduct an experiment to find out.

_____

_____

_____

_____

Open Inquiry

Design an experiment to test the malleability of materials.

Think about how you could test objects for malleability to determine if the objects might be metals.

My question is:

_____

_____

How can I test it?

_____

_____

My results are:

_____

_____

How do acids affect metals and nonmetals?

Materials

- clear plastic cup
- iron nail
- piece of graphite
- vinegar

Purpose

Investigate how acids like vinegar affect materials.

Procedure

1 **Experiment** Pour vinegar into a cup until it is about 3 cm deep. Add a piece of graphite and an iron nail to the vinegar.

2 **Observe** Let the contents of the cup sit for at least 15 minutes. Without moving the cup, look at the graphite and the nail. What do you observe?

_____

_____

Draw Conclusions

3 From your observations, what can you conclude about how an acid affects iron and carbon?

_____

_____

4 **Predict** Magnesium reacts even more with acid. In the box below draw a picture of what you expect to observe when a piece of magnesium is placed in vinegar.

Name _____ Date _____

Hardness vs. Malleability

Materials

- goggles
- metal paper clip
- copper wire

1 ⚠ **Be Careful.** Wear goggles to protect your eyes. Bend one end of a paper clip 90° and then bend it back to its original position. Try the same action with a piece of copper wire.

2 **Predict** How many times can you repeat this step before the paper clip breaks? Before the copper wire breaks? Record how many bends were required to break each.

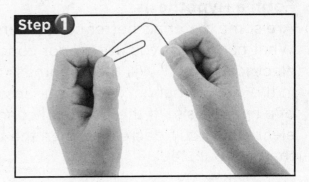

Step **1**

paper clip	copper wire

3 Will the copper wire scratch the paper clip, or will the paper clip scratch the copper wire? Record the results when you try to scratch one metal with the other.

_____

4 **Infer** Which metal was harder? Which was more malleable? Explain your reasoning.

_____

_____

_____

Name _____ Date _____

Structured Inquiry

How can you compare the electrical conductivity of metals and nonmetals?

Form a Hypothesis

Are some materials better conductors than others? What happens if you use a poor conductor in an electrical circuit? Will the brightness of a light bulb in the circuit change? Write your answer in the form of a hypothesis: "If the electrical conductivity in an electrical circuit decreases, then the brightness of the light bulb . . ."

_____

_____

_____

Materials
• battery
• battery holder
• alligator clips
• wire
• miniature bulb
• bulb holder
• copper, aluminum, iron, and tin electrodes 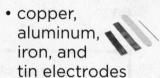

Test Your Hypothesis

1. Place the battery in the battery holder. Connect one alligator clip and wire to one end of the battery holder. Connect another alligator clip and wire to the other end of the battery holder.

2. Connect an alligator clip from the battery to the miniature bulb in the bulb holder. Use a third alligator clip and wire to attach the bulb holder to one end of the copper strip. The copper strip is the material you will test and change—it is your independent variable.

Step 1

© Macmillan/McGraw-Hill

Be a
Scientist

3 **Experiment** Connect the
second wire from the battery
to the other end of the copper
to close the circuit. Observe how
brightly the light bulb glows—this
is your dependent variable.

4 **Observe** Repeat the test for
the other materials. Observe and
record your results.

Step 3

copper	wood	iron	graphite

5 **Classify** Rank the four materials' conductivities
in order from highest to lowest. _____

Draw Conclusions

6 **Infer** Did the light bulb act as a conductivity tester?
Why do you think copper is used to make wire?

_____

_____

_____

_____

_____

7 Do your results support your hypothesis? Explain.

_____

Guided Inquiry

How does combining materials affect conductivity?

Form a Hypothesis

You know how the electrical conductivity of metals and nonmetals compare. How will that property change if different materials are combined? Write your answer in the form of a hypothesis: "If a good conductor is combined with a poor conductor, then the new object's conductivity will be . . ."

_____

_____

Test Your Hypothesis

Design and perform an experiment to determine how conductivity changes when materials are combined. You may want to use steel (a combination of iron, other metals, and carbon) or a pencil (a combination of graphite and wood). Write out the resources you need and the steps you follow. Remember to describe your variables and record your results.

_____

_____

_____

_____

_____

Draw Conclusions

Did your experiment support your hypothesis? Why or why not?

_____

_____

Open Inquiry

Do materials conduct heat energy as well as they conduct electrical energy? Design an experiment to answer your question. Organize your experiment to test only one independent variable, or one item being changed. Write your experiment so that someone else can complete the experiment by following your instructions.

My hypothesis is:

_____

_____

How can I test it?

_____

_____

_____

My conclusions are:

_____

_____

Name _____ Date _____

What happens when ice melts?

Make a Prediction

If you warm ice cubes, they will melt. What happens to the temperature of a cup of ice cubes and water as the ice melts? Write a prediction in the form: "If a cup of ice and water is steadily warmed, then the temperature of the ice water will . . ."

_____

_____

_____

Materials

- plastic or paper cup
- cool water
- ice cubes
- balance
- watch or clock
- thermometer
- heat source (lamp or sunlight)

Test Your Prediction

1 **Measure** Fill a cup halfway with cool water and add four ice cubes.

2 Record the mass of the cup of ice water. Do you think the mass will change as the water warms?

_____

_____

_____

Step **2**

3 **Observe** Swirl the ice and water gently for 15 seconds (don't splash!). Then record the temperature of the mixture. Next, place the cup under the heat source.

Step 3

4 Repeat step 3 every 3 minutes until you have 5 readings after the ice has fully melted.

Temperature
Start

5 Record the mass of the cup of water again.

_____

Draw Conclusions

6 Use your data to make a graph of the temperature of water versus time on another sheet of paper.

7 **Interpret Data** Describe the temperature and mass of the ice water as the ice melted.

_____

_____

_____

Name _____ Date _____

8 Communicate Did your observations support your prediction? Write a report that describes whether or not your prediction was correct.

Explore More

How would the temperature of water change with time as it freezes? Write a hypothesis and design an experiment to test it. Conduct your experiment and report on your findings.

_____

_____

Open Inquiry

As ice changes to water, it stays the same temperature. What about gases? Does water stay the same temperature as it changes to water vapor? Design an experiment to find out.

My question is:

_____

_____

How can I test it?

_____

_____

_____

_____

My results are:

_____

_____

_____

Will ice melt faster in hot water?

Purpose
Determine if ice melts faster in hot water or cold water.

Make a Prediction
People often add ice to hot liquids to make them cooler. But ice melts in cool water, too. In which kind of water, hot or cold, will ice melt fastest? Make a prediction.

_____

_____

Materials
• two ice cubes of the same size
• two clear plastic measuring cups
• hot and cold tap water
• clock

Test Your Prediction

1 Fill one cup about three-fourths full with hot tap water. Fill the other cup with the same amount of cold tap water.

2 Put one ice cube in each cup at the same time. Record the time.

_____

3 Observe both cups as the ice melts. Record the time at which the ice in each cup has completely melted.

Hot: _____   Cold: _____

Draw Conclusions

4 In what order did the ice cubes melt?

_____

_____

5 What is the relationship between the temperature of the surroundings and the time it takes ice to melt?

_____

_____

Name _____ Date _____

Changing Balloons

Materials
- balloons
- string
- pail or other container
- cold water
- ice

1 **Predict** What will happen to the volume of a balloon filled with warm air as it is cooled? Record your prediction.

_____

2 Blow up a balloon. Air from your lungs is warm. Tie off the balloon and measure it around with a piece of string.

_____

3 Submerge the balloon in a pail of ice water for a few minutes. Measure it again with the piece of string. Record your observations.

_____

_____

4 **Infer** How does the motion of molecules explain what you observed? Write out your ideas

_____

_____

_____

_____

Use Variables

Particles in hot liquid move faster than those in cold. Since hot water has more energy to get rid of before it freezes, it shouldn't freeze as fast as cold water.

That's what many people thought. But scientists wanted to know for sure, so they did a series of experiments and recorded their observations. Those experiments changed only one thing at a time. That way the scientists knew what caused the outcome they observed. What they changed is called the independent **variable**. They learned that sometimes hot water freezes faster than cold water—this is called the Mpemba effect.

▶ **Learn It**

When you use variables, you change one thing at at time to see how it affects the outcome of the experiment. The thing you change is the independent **variable**. The outcome is the dependent variable. The way the dependent variable changes depends on the way the independent variable changes.

For this experiment the independent variable is the starting temperature of the water. The time it takes the water to freeze is the dependent variable. You will change the starting temperature of the water and record how this affects the time it takes the water to freeze.

Name _____   Date _____

▶ **Try It**

1 Use a chart to record your data.

2 Fill a cup with 120 mL of hot water. Label it *Hot Water*. Fill a cup with 120 mL of cold water. Label it *Cold Water*. Fill a cup with 80 mL cold and 40 mL hot water. Label it *Cool Water*. Fill a cup with 80 mL hot and 40 mL cold water. Label it *Warm Water*.

3 Record the temperature of each cup of water in your chart. This is the independent **variable**.

Materials

- hot water
- cool water
- plastic cups
- thermometer
- graduated cylinder
- labels
- freezer

	Time to Freeze		
	Temperature	**Starts freezing**	**Ends freezing**
Hot water			
Warm water			
Cool water			
Cold water			

4 Place all the cups in a freezer at the same time. The cups should be close together and on the same level.

5 Check the freezer every 10 minutes. Record when the water in each cup starts to freeze. Record when the water in each cup is completely frozen. These are both dependent variables.

© Macmillan/McGraw-Hill

► **Apply It**

1 Which water froze first: cold, cool, warm, or hot water? Repeat the experiment to confirm your findings.

_____

2 Scientists changed the independent **variable** to learn about the Mpemba effect. What did you learn from your results? Do you agree that the Mpemba effect is real?

_____

3 What do you think would happen if you used really icy or even hotter water? Are you still changing the same independent **variable**? Try it and record data about the investigation. Use that data to help you develop an opinion about how water freezes.

	Time to Freeze		
	Temperature	Starts freezing	Ends freezing
Hot water			
Warm water			
Cool water			
Cold water			

_____

_____

© Macmillan/McGraw-Hill

Name _____ Date _____

How can you speed up mixing?

Make a Prediction

Which do you think will speed up the mixing of sugar in water more: crushing the sugar, stirring the water, or heating the water? Record your prediction.

_____

_____

Materials

- sugar cubes
- cold and hot water
- plastic cups
- stop watch
- spoon

Test Your Prediction

1 Use a table to record your observations.

Step 1

	Water	Temperature	Sugar	Time to Dissolve
Control	½ cup	Cold	Whole	
Test 1	½ cup	Cold	Crushed	
Test 2	½ cup	Cold	Stirred	
Test 3	½ cup	Hot	Whole	

	Water	Temperature	Sugar	Time to Dissolve
Control				
Test 1				
Test 2				
Test 3				

2 **Experiment** Take 1 whole sugar cube and place it in 1/2 cup of cold water. Record the time it takes to completely dissolve. This is the control mixing time.

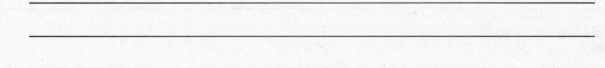

Step **2**

3 Repeat step 2 using 1 crushed cube, then with 1 cube while stirring the water, then with 1 cube in 1/2 cup of hot water.

Draw Conclusions

4 **Interpret Data** Look at your table. Which method had the shortest mixing time? Was it close to or very different from the control?

_____

_____

_____

5 **Infer** How do you think you could create the shortest mixing time possible? Write a report and justify your answer.

_____

_____

_____

Explore More

Do you think there are any other methods that could decrease the mixing time? Design an experiment that could provide information and test your prediction. Carry out the experiment and record your results.

_____

_____

_____

_____

Open Inquiry

What do you think would happen to the mixing time if ice water were used instead of heated water? Think of your own question about the effect of ice water on mixing time. Make a plan and carry out an experiment to answer your question.

My question is:

_____

_____

How can I test it?

_____

_____

_____

My results are:

_____

_____

How can you speed up the making of tea?

Materials

- two foam cups
- cold tap water
- hot tap water
- two tea bags

Purpose
Determine the conditions that help you make tea faster.

Make a Prediction
People make tea by putting the leaves of tea plants in water. Substances in the tea leaves mix with the water to form the tea. What conditions will help you make tea the fastest? Make a prediction.

_____

Test Your Prediction

1. Fill one foam cup about three-fourths full with cold tap water. Fill the other cup with the same amount of hot tap water.

2. Gently lower a tea bag into each cup at the same time. Allow the tea bags to float undisturbed on the water. After 2 or 3 minutes, record what you see.

_____

_____

3. Dip the tea bags in the cups two or three times. What happens in each cup?

_____

Draw Conclusions

4. What temperature of water makes tea form fastest?

_____

5. What conditions speed up the making of tea?

_____

_____

Name _____ Date _____

Temperature in Solutions

Materials

- plastic cups
- spoons
- sugar
- hot and cold water

1 **Predict** Do you think you could dissolve more sugar in hot water or cold water? Why? Write down your reasons

_____

_____

_____

2 **Observe** Fill a cup with cold water. A level spoonful of sugar is about 28 g. Record how many grams of sugar will dissolve in the water as you stir. Repeat with hot water.

	Cold water	Hot water
Grams of sugar		

3 Which water dissolved the most sugar? How could you tell?

_____

_____

4 Was your prediction correct? Write out your findings.

_____

_____

_____

Structured Inquiry

How can you separate mixtures?

Form a Hypothesis

Are all mixtures made in the same way? Will different methods of separation work equally well on the same mixture? Write your answer in the form of a hypothesis "If the method for separating a mixture changes, then . . ."

_____

_____

_____

Materials

- sand
- gravel
- two bowls
- spoon
- iron filings
- sieve
- tweezers
- bar magnet

Test Your Hypothesis

1 Take a cup of sand and gravel and pour it into a bowl. Add a spoonful of iron fillings and mix them into the sand and gravel.

2 **Experiment** For about one minute, use a sieve to try and separate the mixture into another bowl. Record how well the mixture separated—the dependent variable for this experiment.

_____

_____

Step **2**

Name _____ Date _____

Test Your Hypothesis

3 Remix the ingredients. For about one minute use the tweezers to try and separate the mixture. Record your results.

_____

_____

Step **3**

4 Repeat step 3 using a magnet.

_____

_____

Draw Conclusions

5 **Use Variables** What was the independent variable for this experiment? Were there controlled variables?

_____

_____

_____

Step **4**

6 **Interpret Data** Rank the separation methods from least to most effective. Make sure to give reasons for your ranking.

_____

_____

_____

7 Did your results support your hypothesis? Write a report explaining why or why not.

Guided Inquiry

How can water separate a mixture?

Form a Hypothesis

You have seen how a mixture's properties affect how you can separate a mixture. Adding water to a mixture changes the properties of that mixture. How would this change the way you could separate a mixture of salt, sand, and sawdust? Write your answer as a hypothesis in the form, "If water is added to a mixture of salt, sand, and sawdust, then the best method to separate the mixture will . . ."

_____

_____

Test Your Hypothesis

Briefly try to separate a mixture of salt, sand, and sawdust using only a filter. Next, design a procedure that uses both water and a filter to completely separate the mixture into three piles: sand, salt, and sawdust. Write out the resources and steps you would follow. Record your variables, results, and observations as you follow your plan.

My Plan and Results
_____
_____
_____
_____
_____
_____
_____

© Macmillan/McGraw-Hill

Draw Conclusions

Did your experiment support your hypothesis? Why or why not?

_____

_____

Open Inquiry

For example, how can you separate black ink into different-colored inks? Design an experiment using chromatography paper to answer your question. Your experiment must be written so that someone else can complete the experiment by following your instructions.

My hypothesis is:

_____

How can I test it?

_____

_____

My conclusions are:

_____

_____

_____

Does mass change in a chemical change?

Form a Hypothesis

Does the total mass of matter change when one substance turns into another? Think about chemical changes you have observed: an egg being cooked or wood burning in the fireplace. Write your answer as a hypothesis in the form: "If a chemical reaction occurs, then the total mass . . ."

_____

_____

Materials

- safety goggles
- washing soda solution (sodium carbonate)
- sealable bag
- Epsom salt solution (hydrated magnesium sulfate)
- small paper cup
- equal pan balance

Test Your Hypothesis

1 ⚠ **Be Careful.** Wear safety goggles! Pour 40 mL of washing soda solution into a bag. Place 40 mL of Epsom salt solution in a paper cup. Put the cup inside the bag so that it rests upright. Seal the bag.

2 **Measure** Place the bag on a balance. Don't mix the solutions! Record the mass—it is the dependent variable for this experiment.

Step **2**

_____

© Macmillan/McGraw-Hill

3 **Observe** Without opening the bag, pour the solution in the cup into the solution in the bag to cause a chemical change.

4 Record the mass of the bag and its contents.

_____

Draw Conclusions

5 What is the independent variable in this experiment? Were there other variables you controlled?

_____

_____

_____

6 **Interpret Data** How did mass change during the chemical reaction?

_____

7 Does the data support your hypothesis? If not, how would you change your hypothesis?

_____

_____

Explore More

Do you think that volume is conserved in a chemical change?
Plan an experiment that would provide information to support
your conclusion.

_____

_____

_____

Open Inquiry

What do you think would happen to the mass of another
substance that was having a chemical reaction? Will mass be
conserved with different kinds of matter?

Think of your own question about chemical reactions and
mass. Make a plan and carry out an experiment to answer
your question.

My question is:

_____

_____

_____

How can I test it?

_____

_____

My results are:

_____

Name _____ Date _____

What are some signs of a chemical reaction?

Materials
• lemon juice
• baking soda
• foam cup
• thermometer
• spoon

Form a Hypothesis

Lemon juice and baking soda are common substances. Each has its own set of physical properties. What would happen if you mixed the two together? Would you be able to tell if a chemical reaction is happening? Form a hypothesis.

_____

_____

Test Your Hypothesis

1 Pour lemon juice into a foam cup until the cup is about half full.

2 Measure the temperature of the lemon juice. What was the temperature?

_____

3 Stir one teaspoon of baking soda into the lemon juice. What, if any, changes did you observe?

_____

4 Measure the temperature of the solution in the cup. What was the temperature this time?

_____

Draw Conclusions

5 What happened to the temperature as the chemical change took place?

_____

6 What are two signs that a chemical change has happened?

_____

Chemical Cents

Materials
• vinegar
• salt
• penny
• cup

1. Pennies are coated in copper, which corrodes easily. Find a dull and tarnished penny.

2. **Observe** Put the penny in a cup of salt and vinegar. Record your observations.

   _____

3. Are there any signs of a chemical reaction? Take the penny out and let it dry in the air. Do any more chemical reactions occur? How do you know?

   _____

   _____

   _____

4. You can also place a steel nail or screw in the vinegar after removing the penny. The copper oxide that was removed by the vinegar will bond to, or plate, the steel. Hydrogen gas will also bubble up. If you have difficulty getting the copper to plate the steel, use several pennies. Try this and record your results.

   _____

   _____

Name _____ Date _____

Which are acids and which are bases?

Purpose

Acids turn blue litmus paper red and have no effect on red litmus paper. Bases turn red litmus paper blue and have no effect on blue litmus paper. You will examine household solutions to see if they are acids or bases.

Procedure

1 **Predict** Use your sight and sense of smell to predict whether each household solution is an acid or base.

Materials

- goggles
- gloves
- apron
- red and blue litmus papers
- samples of household solutions
- pH paper

	Predict: Acid or Base	Effect on Red Litmus	Effect on Blue Litmus	pH strip	Result: Acid or Base
Sample # 1:					
Sample # 2:					
Sample # 3:					
Sample # 4:					

Use with **Lesson 4**
Acids, Bases, and Salts

2 ⚠ **Be Careful.** Wear goggles, gloves, and an apron! Record what happens as you dip a strip of each color of litmus paper into sample #1.

3 **Classify** Repeat step 2 with other solutions your teacher has provided and record your results.

Step **3**

4 Now test each solution with a small strip of pH paper. pH is a measure of acidity. Use the color scale provided to find and record the pH on your chart.

_____

_____

_____

_____

Draw Conclusions

5 Organize the samples you tested into acids and bases. What does pH tell you about acids and bases?

_____

_____

6 **Communicate** Do all the acids or all the bases have anything in common? Do any seem more acidic or more basic than others? Why do you think so? Write down your reasons in a report.

Name _____ Date _____

Explore More

Some people claim that cola drinks are very acidic and can dissolve iron nails overnight! Design and perform an activity in which you test the acidity of colas. Are they really strong acids?

_____

_____

Open Inquiry

Think of your own question about the acids and bases you have in your house. Make a plan and carry out an experiment to answer your question.

My question is:

_____

How can I test it?

_____

_____

My results are:

_____

_____

What are some natural acid-base indicators?

Materials

- goggles
- apron
- blueberries
- beet juice
- four paper cups
- lemon juice
- liquid soap
- clock

Purpose

Discover whether foods can be acid-base indicators.

Procedure

1. Put one blueberry, or a small amount of blueberry jam or jelly, in each of two paper cups. Pour a small amount of beet juice into the two other cups.

2. Pour a small amount of lemon juice, enough to cover the blueberry, in one cup. What color is the juice after about 5 minutes?

_____

3. Pour a small amount of liquid soap, enough to cover the blueberry, in the other cup. What color is the soap after about 5 minutes?

_____

4. Repeat steps 2 and 3 with the cups that contain beet juice. You will need to use more lemon juice and soap this time. What color is the liquid in each cup after 5 minutes?

_____

_____

Draw Conclusions

5. Can either of the foods tested be used as acid-base indicators? Which ones?

_____

6. What color indicates an acid? What color indicates a base?

_____

© Macmillan/McGraw-Hill

Name _____ Date _____

Indicating Ink

Materials
• cotton swab
• baking soda solution
• grape juice

1 Dip a cotton swab in baking soda solution. Use it to write a message to a partner on a piece of paper.

2 Allow the paper to dry. Then switch papers with your partner.

3 **Observe** Is the message invisible? Use another swab to "paint" the paper with grape juice. Record your observations.

_____

_____

4 **Infer** Is the grape juice an acid-base indicator? Why or why not?

_____

_____

5 What are some other natural acid-base indicators?

_____

_____

_____

_____

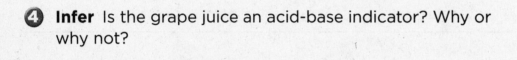

Name _____ Date _____

How is speed measured?

Form a Hypothesis

How do you think speed depends on the distance an object travels? Write your answer as a hypothesis in the form "If the distance a marble travels increases, then . . ."

_____

_____

_____

Materials

- marble
- index card
- masking tape
- meterstick
- stopwatch

Test Your Hypothesis

1 Make a marble launcher out of an index card. Use the pattern provided. Place the launcher on a long, flat, and smooth surface.

2 Place a piece of tape in front of the launcher—this is your starting point. Use a meterstick to place a piece of tape one meter from your starting point. This is your "finish line" and your independent variable.

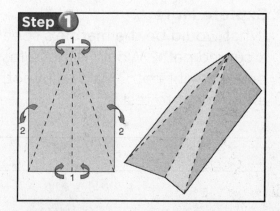

3 **Measure** Roll a marble down the launcher. Use a stopwatch to time it as it travels from the starting point to the finish line. Repeat twice more and calculate an average time—this is your dependent variable. Record your results and the average time on another sheet of paper.

4 Repeat step 3 for finish lines at 2 m and 3 m. Record your results.

© Macmillan/McGraw-Hill

Name _____ Date _____

Draw Conclusions

⑤ Use Numbers Divide each distance by its average time. This value is the average speed of the marble over that distance.

⑥ Communicate Did the launcher roll the marble at the same speed every time? Write a report describing the motion of the marble out of the launcher.

_____

_____

_____

_____

Explore More

What would be the marble's speed if it traveled on a curved path? Would it move faster or slower than on a straight path? Write a hypothesis and design an experiment to test it.

_____

_____

Open Inquiry

How does the slope of the launcher affect the speed of a launched marble? Think of your own question about the effect of slope steepness on speed and then design and carry out an experiment to test it.

My question is:

_____

_____

How I can test it:

_____

_____

_____

My results are:

_____

_____

_____

_____

Name _____ Date _____

How fast will it sink?

Purpose

Determine the speed at which thumbtacks and paper clips sink in water.

Procedure

1 Fill the soda bottle with water

2 Measure and record the height of the water in the soda bottle in centimeters.

Height: _____

3 Have a timer ready to start just when an object is dropped into the soda bottle.

4 Drop a paper clip into the water. Start timing when the paper clip first touches the water until it reaches the bottom of the bottle. Record the time in seconds in a table on a separate piece of paper.

5 Repeat step 4 with the other two paperclips and the 3 thumbtacks.

Draw Conclusions

6 How can you determine the average speed of a sinking object?

_____

_____

_____

7 What were the average speeds of the sinking objects?

_____

_____

Materials

- 3 thumbtacks
- 3 paper clips
- 2-liter soda bottle
- water
- ruler
- timer or stopwatch

Name _____ Date _____

Stepping Speed

Materials

• timer

① Organize into teams of three: a quick-stepper, a timer, and a measurer.

② **Measure** At the command "Go!", the stepper takes two quick steps, covering as much distance as possible. The timer records the time and the measurer records the distance. The stepper then repeats this process for three, four, and five quick steps.

③ Repeat the set of measurements two more times with everyone changing roles.

④ Make a line graph of your data for distance and time on a separate piece of paper. Distance should go on the vertical axis and time on the horizontal axis.

⑤ **Interpret Data** Does distance change steadily with time in your graph? Why or why not?

_____

_____

_____

_____

Name _____ Date _____

Do heavier objects fall faster?

Materials

- equal pan balance
- golf ball
- tennis ball
- cotton ball

Form a Hypothesis

In the late 1500s, Galileo argued that weight should not affect how fast something falls. Do you agree? Write a hypothesis in the form "If the mass of an object increases, then . . ."

_____

_____

Test Your Hypothesis

1 **Observe** Use a balance and standard masses to determine the mass of each object. Order the objects from lightest to heaviest and write down your list.

_____

_____

_____

_____

2 **Experiment** Hold two of the objects at the same height in front of you. Drop them at exactly the same time. Record which object hits first or if they hit at the same time. Repeat two more times to verify your result.

_____

_____

_____

3 Repeat step 2 until you've tested all possible pairs of objects.

Draw Conclusions

4 **Interpret Data** Was your hypothesis correct? Write a brief explanation of your answer.

_____

_____

_____

_____

5 **Infer** In your experiment, the objects were falling through air. There is no air on the Moon. How would the falling rate of a tennis ball and a cotton ball compare on the Moon? Why?

_____

_____

_____

Explore More

How would the results of this experiment change if you dropped objects with the same mass, but different densities? Write out a hypothesis. Then use balloons at different levels of inflation to test your hypothesis. Write a summary of your results.

_____

_____

_____

_____

_____

Name _____ Date _____

Open Inquiry

Form a hypothesis about the speeds of falling objects
that have different surface areas. Then design and carry
out an experiment to test it.

My question is:

_____

_____

_____

How I can test it:

_____

_____

_____

_____

My results are:

_____

_____

_____

_____

Why does the stack of washers stay in place?

Purpose
Demonstrate a force acting on a stack of washers.

Materials
- 8 large metal washers
- craftstick
- smooth desk top

Procedure

1 Place a stack of 8 washers near the edge of a smooth desk or table.

2 Hold the craftstick so that it is flat against the desk and almost touching the bottom washer.

3 Bring the craftstick back away from the washers about 10 cm.

4 **Experiment** Being sure to keep the craftstick parallel to the table, use a quick flick to strike only the bottom washer.

5 Repeat step 4 two more times.

6 **Observe** What happened to the bottom washer each time it was hit with the craftstick?

_____

Draw Conclusions

7 Why was only the bottom washer pushed out of the stack?

_____

_____

_____

_____

© Macmillan/McGraw-Hill

Name _____ Date _____

Unbalanced Balloon Force

Materials

• balloon

• string

1. Pass thread or string through two short lengths of soda straw. Then stretch the string tightly between two chairs.

2. Inflate a balloon. Hold the neck closed while you tape it to the straws.

Step 2

3. **Observe** Let go of the balloon's neck and record what happens.

_____

_____

_____

4. **Infer** Did an unbalanced force act on the balloon? Explain.

_____

_____

_____

5. How would the balloon motion change if you inflate the balloon more than before? Write down your prediction and experiment to test it. Report what you discover.

_____

_____

Measure

If an unbalanced force acts on an object, that object will accelerate. As the force changes, how will the acceleration change? Will an increasing force increase the final velocity of an object? Scientists often ask themselves questions they don't know the answer to. To find the answers, they **measure** and observe the things around them.

▶ **Learn It**

When you observe you use one or more of your senses to identify or learn about an object. When you **measure** you find the size, distance, time, velocity, mass, weight, or temperature.

It is important to record measurements and observations you make during your experiment. Once you have enough information, you can make predictions about what might happen if you changed a variable in the experiment.

© Macmillan/McGraw-Hill

Name _____ Date _____

Measure

• Materials •

▶ **Try It**

Newton's second law of motion ($F = m \times a$) tells you how unbalanced forces, mass, and acceleration relate.

You can launch a model car with rubber bands and then record its final speed. How will the car's speed depend on the number of rubber bands used? Write a hypothesis in the form "If the number of rubber bands used to launch a model car increases, then . . ."

Materials

- model car
- masking tape, rubber bands
- 2 wood blocks
- thumbtacks
- meter stick
- safety goggles
- stopwatch

_____

_____

_____

1 Place a 10 cm strip of masking tape on the floor. Hold the wood blocks on either side of the tape. Stretch a rubber band between the thumbtacks.

2 Tape a "finish line" on the floor 1.5 m (59 in.) from the rubber band.

3 ⚠ **Be Careful.** Wear goggles. Pull a toy car back against the rubber band a few centimeters. Mark this starting point with a small piece of tape and use it for the rest of the experiment.

4 **Measure** Launch the car and use a stopwatch to time how long it takes for the car to travel 1.5 m. Repeat this procedure two more times and find the average time. Record this on the table on the next page.

5 Repeat step 4 with two rubber bands.

	Time 1	Time 2	Time 3	Average Time	Average Speed
One Rubber Band					
Two Rubber Bands					

► **Apply It**

Now use the data you **measured** and recorded to answer questions.

1 Divide the distance (1.5 m) by the average time for one rubber band to get the car's speed. Repeat for the trials with two rubber bands. Record this on the table above.

2 According to the speeds you found, was your hypothesis correct? Explain.

_____

_____

_____

_____

3 Can you predict how the car's speed would change if you were to tape a second car on top of it? How do Newton's second law and your experiments support this prediction?

_____

_____

_____

_____

Name _____ Date _____

What happens to energy?

Form a Hypothesis

The energy an object has changes as it is pulled by gravity. What will happen if you let a marble roll in a bicycle tire? Write a hypothesis in the form "If the height a marble is released from is increased, then . . ."

_____

_____

_____

Materials

- section of old bicycle tire (or piece of split garden hose)
- masking tape
- marble

Test Your Hypothesis

1. You will need to work in a group. One member of your group should hold the section of tire firmly on a tabletop. Use a piece of masking tape to mark a starting point on one side of the tire.

2. **Observe** Release the marble at the starting point and let it roll in the tire. Observe what the marble does until it comes to a stop. The marble's actions are your dependent variable. Repeat several times to verify your observations.

3. Repeat steps 1 and 2 for two more starting points. They should be at different heights. The height of the marble is your independent variable.

Step 2

© Macmillan/McGraw-Hill

Draw Conclusions

4 **Interpret Data** According to your observations, was your hypothesis correct? Explain.

_____

_____

5 **Infer** When did the marble move the fastest? Did it have more or less energy there than when it began? How do you know?

_____

_____

_____

Explore More

Why did the marble eventually stop? What effect did the texture of the inside of the bicycle tire have? Write a hypothesis and design an experiment to test it.

_____

_____

_____

_____

_____

Open Inquiry

A stretched rubber band has energy of position. Form a hypothesis about energy of motion, then design an experiment to measure the energy.

My question is:

_____

_____

How I can test it:

_____

_____

_____

My results are:

_____

_____

_____

_____

Is it potential or kinetic?

Materials
• yo-yo

Purpose

Demonstrate the transfer of kinetic and potential energy.

Procedure

1 Wind the string around the axle of the yo-yo. Hold the yo-yo in your hand.

2 Extend your hand away from your body.

3 **Observe** Holding the string, drop the yo-yo. Allow the yo-yo to drop without pulling back on the string.

4 **Observe** Rewind the yo-yo and drop it again, this time snapping your wrist as the yo-yo reaches the bottom of the string.

Draw Conclusions

5 At which point did the yo-yo have the most potential energy?

_____

_____

6 At which point did the yo-yo have the most kinetic energy? How do you know?

_____

_____

7 What caused the yo-yo to return when you snapped the string?

_____

_____

Name _____ Date _____

Measuring Used Energy

Materials
- book
- spring scale
- string

1 Tie a loop of string around a book. Hook the string to a spring scale.

2 **Measure** Slide the book across a table by pulling on the spring scale. Keep the force reading on the spring scale constant. Record your results.

3 Hang the book from the spring scale. Record the book's weight.

4 Which is more work: sliding the book for 1 meter or lifting the book a distance of 1 meter? Explain.

_____

_____

_____

5 **Infer** If you lift the book up to a certain height, it gains potential energy. When you slide the book a given distance, it is neither lifted nor left with kinetic energy. Where does the energy from the work go when you slide the book?

_____

_____

_____

What makes work easier?

Make a Prediction

Will pulling a toy car up a ramp be more work than lifting it straight up? Write a prediction stating whether you think that pulling the car up the ramp will be more or less work than pulling it straight up.

_____

_____

_____

_____

Materials

- books
- ruler
- toy car
- spring scale

Test your Prediction

1 Hang the toy car from the spring scale and read its weight in newtons. Record your results.

_____

Step 1

2 Use 4 books to build a ramp as shown. Measure the height of the ramp using the ruler. Pull the toy car up the ramp at a steady speed. Read the force required in newtons. Measure the distance the car traveled along the ramp using a ruler. Record your results.

_____

3 Repeat your measurements to verify your results.

Step 2

Draw Conclusions

4 **Use Numbers** Calculate the work it takes to lift the car to the height of the ramp and the work it takes to pull the car up the ramp. Remember Work = Force × Distance. Were your predictions correct? Explain.

_____

_____

_____

5 **Communicate** To get something you want, you usually have to pay for it. What "price" do you pay when you use a ramp to help lift something?

_____

_____

6 **Infer** Are there any additional forces acting on the car when you use the ramp? How might these forces affect the work you did?

_____

_____

_____

Explore More

How would changing the angle of the ramp affect the pulling force? Make a prediction and design an experiment to test it. Perform the experiment to check to see if your prediction was correct.

_____

_____

_____

Open Inquiry

What simple materials might be used to help reduce the friction on a ramp? Design and carry out an experiment to answer this question.

My question is:

_____

_____

How I can test it:

_____

_____

_____

My results are:

_____

_____

Name _____ Date _____

How simple is the machine?

Purpose
Demonstrate that the screw is a simple machine.

Procedure

1. Using only your fingers, put the nail into the piece of cardboard.

2. Using only your fingers, put the screw into the piece of cardboard.

3. Wind the paper around the pencil so that it looks like a screw.

4. **Observe** What is the difference in the shapes of the nail and the screw?

   _____

   _____

5. **Observe** What was the shape of the piece of paper before you wrapped it around the pencil? What simple machine is a screw?

   _____

   _____

Draw Conclusions

6. What type of simple machine is at the tip of both the nail and the screw?

   _____

Materials

- sheet of paper folded diagonally and cut to be about 8" X 11" on the sides and 13.6" along the hypotenuse

- pencil

- nail

- screw

- cardboard

© Macmillan/McGraw-Hill

Levers and Effort

Materials
• meterstick
• string
• spring scale
• paper clips
• weight

1 Hang a meterstick by its center so that it balances.

2 Use a paper-clip hook to hang a weight 25 cm from the middle of the meterstick. Position a second hook on the other side, 25 cm from the middle. Attach a spring scale to this hook and measure the downward force it takes to hold the bar in a level position. Record your reading.

3 **Repeat** with the spring scale positioned at 15 cm and 35 cm. Record your readings.

_____

_____

_____

4 **Interpret Data** In each case, the resistance arm was 25 cm long. How was the length of the effort arm (the spring-scale arm) related to the force needed to keep the meter stick level? Explain.

_____

_____

_____

_____

© Macmillan/McGraw-Hill

Name _____ Date _____

Which can give you more heat?

Form a Hypothesis

If you mix the same amount of water or oil into ice water, which one will warm the ice water more? Write your answer as a hypothesis in the form "If the same amount of room-temperature water and oil is added to ice water, then . . ."

_____

_____

_____

_____

Materials

- ice
- water
- graduated cylinder
- plastic cups
- cooking (corn) oil
- thermometer

Test Your Hypothesis

1. Pour 100 mL of ice water (but don't include any ice) into 2 cups. Pour 100 mL of room-temperature water and cooking oil into 2 different cups. Record the temperature of each.

2. **Experiment** Mix a cup of ice water into the cup of room-temperature water and stir for 2 minutes. Record its temperature. Repeat this process for the cooking oil.

Step 1

Step 2

Water

Use with **Lesson 1**
Heat

© Macmillan/McGraw-Hill

3 **Use Numbers** Subtract the starting temperature of the ice water from the final temperature of each mixture. This gives you the temperature change of the ice water for each experiment.

	Starting Temperature	Final Temperature	Temperature Change
Water			
Cooking Oil			

Draw Conclusions

4 **Interpret Data** How do the temperature changes compare? Was your hypothesis correct? Explain.

_____

_____

_____

5 **Infer** Based on your answers in step 4, is heat the same thing as temperature? Explain.

_____

_____

_____

_____

Name _____ Date _____

Explore More
Which would cool faster starting at the same high temperature, 100 mL of cooking oil or 100 mL of water? Write a hypothesis. Then design and carry out an experiment to test it.

_____

_____

_____

_____

Open Inquiry
You used liquids to add heat to ice water. What solids add more heat to ice, causing it to melt? Think of a question about how two different solids at the same temperature will affect melting ice. Make a plan and carry out an experiment to answer your question.

My question is:

_____

_____

How I can test it:

_____

_____

_____

_____

My results are:

_____

_____

_____

_____

What can add more heat to ice?

Form a Hypothesis

You add ice to a drink to make it colder. But how does this happen? And why does the ice always melt? Remember, heat moves from a warmer substance to a colder substance. That means your drink is making the ice warmer, instead of the ice making your drink colder. Do some liquids with the same temperature have more heat to give to the ice and make it melt? Develop a hypothesis you can test with an experiment that uses the materials above.

Materials

• 2 clear plastic cups

• water at room temperature

• cooking oil at room temperature

• 2 ice cubes

_____

_____

Test Your Hypothesis

1 Put equal amounts of room temperature water and cooking oil in separate plastic cups.

2 Place one ice cube in each cup at the same time.

3 **Observe** Watch the ice cubes for several minutes. Describe what you observe.

_____

_____

Draw Conclusions

4 Did one ice cube completely melt sooner than the other? Explain.

_____

5 **Infer** Which substance contains the most thermal energy that can be transferred as heat?

_____

Name _____ Date _____

Thermal Differences

Materials

• sand

• water

• lamp

• cups

• thermometers

1 Fill one cup with water and the other with sand. Place a thermometer in each material. Record the temperatures.

2 **Predict** Which material do you think will heat more rapidly when placed under a lamp?

_____

_____

3 Arrange a lamp so that it shines evenly on both cups. Every minute for 10 minutes, measure and record the temperatures.

4 Graph your data for each material as temperature versus time on a separate piece of graph paper.

5 Was your prediction correct? How do you know?

_____

_____

_____

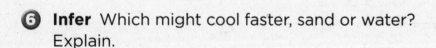

6 **Infer** Which might cool faster, sand or water? Explain.

_____

_____

_____

Form a Hypothesis

You know that heat flows from warmer to cooler objects until they both reach the same temperature. How is that temperature affected by the amount of each object?

Scientists use observations or theories to help them **form a hypothesis**. When you form a hypothesis, you make a testable statement about what you think is logically true.

▶ **Learn It**

A hypothesis is a statement about the effect of one variable on another. It should be based on observations or collected data. For example, when you drink hot chocolate, you might notice that it cools faster when you add ice to it. Based on this observation, you might **form a hypothesis** like "If increasingly colder substances are added to hot chocolate, then it will cool faster."

A hypothesis is tested by conducting an experiment. In this experiment, you will test how much hot water cools when room-temperature water is added. Think about observations you have made in the past involving temperature changes. Write a hypothesis in the form "If increasingly larger amounts of room-temperature water are added to hot water, then . . ."

_____

_____

_____

Name _____ Date _____

▶ **Try It**

In this activity, you will observe how water temperatures change in order to test your **hypothesis**.

Materials

- graduated cylinder
- hot and room-temperature water
- cups
- thermometer
- stopwatch

❶ Use the graduated cylinder to pour 25 mL of room-temperature water into one cup and record its temperature in the chart.

Water Added	Added Water's Temperature	Hot Water's Starting Temperature	Hot Water's Ending Temperature	Hot Water's Temperature Change
25 mL				
50 mL				
75 mL				
100 mL				

❷ Pour 75 mL of hot water into a different cup and record its temperature on the chart.

❸ Add the room-temperature water to the hot water and start the timer on the stopwatch. Place a thermometer in the water and observe its temperature after 2 minutes. Record the new temperature of the hot water.

❹ Repeat steps 1 to 3 with 50 mL, 75 mL, and 100 mL of room-temperature water. You are changing this variable in order to test your hypothesis.

© Macmillan/McGraw-Hill

▶ **Apply It**

1. Subtract the final temperature of the hot water from its starting temperature for each trial. Record your results on your chart.

2. Use the data in your chart to form a graph. On the horizontal axis plot the amount of room-temperature water added to the hot water. On the vertical axis plot the change in temperature of the hot water.

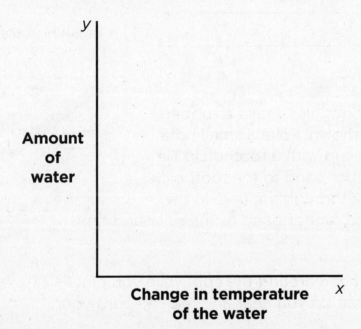

Amount of water

Change in temperature of the water

3. Was your **hypothesis** correct? How do you know?

_____

_____

4. Did the results of the first three trials make it easier to understand what would happen the last time? Why or why not?

_____

_____

_____

Name _____ Date _____

What makes sound?

Form a Hypothesis

When you pluck the rubber band on the "instrument" shown, it makes sound. How will this sound depend on the way you pluck the rubber band? Write your answer as a hypothesis in the form "If the rubber band is plucked with increasing force, then the sound . . ."

_____

_____

Materials

- goggles
- paper cup
- toothpick
- rubber band
- wood or plastic ruler
- masking tape

Test Your Hypothesis

1. ⚠ **Be Careful.** Wear goggles. Make a rubber-band instrument as shown. Poke a small hole in the bottom of the cup with a toothpick. Tie one end of a cut rubber band to the toothpick. Thread the toothpick through the hole in the cup. Tie the stretched rubber band to the ruler and tape the ruler to the cup.

2. **Observe** Wrap one hand around the cup while you pluck the rubber band. What do you hear and feel? Record your observations.

_____

Step 1

Step 2

Use with **Lesson 2**
Sound

3 Pluck the rubber band both gently and forcefully. Record how the sound is affected. Repeat your actions to verify your results.

_____

_____

Draw Conclusions

4 **Interpret Data** Based on your observations, was your hypothesis correct? Explain.

_____

_____

5 **Infer** How do you think your rubber-band "instrument" made sound? Use your observations from step 2 to help you.

_____

_____

Explore More

How will stretching the plucked rubber band affect whether the pitch is high or low? Write out your hypothesis. Then carry out experiments to test it.

_____

_____

_____

_____

Name _____ Date _____

Open Inquiry

What causes the pitch of a wind instrument such as a trumpet to be high or low? Think of a question about how to make different sounds by using empty soda bottles. Make a plan and carry out an experiment to answer your question.

My question is:

_____

How I can test it?

_____

_____

_____

My results are:

_____

_____

_____

How does a drum make sound?

Form a Hypothesis

A drum is an instrument that is played by striking it. But what makes the drum's sound? How can the sound change? Write a hypothesis. Start with "If I stike the head of a drum gently, it will . . ."

_____

_____

Materials

- empty coffee can
- sheet of paper
- rubber band
- paper clips
- pencil

Test Your Hypothesis

1. Make a drum by wrapping a piece of paper over the open top of a large can. Secure the paper snugly in place with a rubber band.

2. Place a few paper clips on the drum head.

3. Lightly tap the drum head with the eraser end of the pencil. What do you observe?

_____

4. Tap the drum head more vigorously with the eraser end of the pencil. What do you observe?

_____

Draw Conclusions

5. What was vibrating to produce the sound?

_____

6. What did the paper clips show?

_____

7. What medium carried the sound to your ears?

_____

© Macmillan/McGraw-Hill

Name _____ Date _____

Sound Carriers

1 **Predict** Will you be able to hear the sound from a radio better through air, water, or wood?

_____

_____

2 Put a radio on a wooden table. Put your ear on the other side of the table and listen. Now lift your head. Record your observations.

	Against object	Away from object	Rating
Air			
Water			
Table			

3 Fill a plastic bag with water. Hold the bag against your ear. Then hold the radio against the other side of the bag. How loud is the radio? Move your ear away from the bag. How loud is the radio now? Record your observations.

4 Rate wood, air, and water as sound mediums from worst to best.

5 **Infer** Foam is less dense than wood or water, but more dense than air. How well do you think foam will carry sound?

_____

_____

Structured Inquiry

How can you change a sound?

Materials

- scissors
- 10 straws
- ruler
- masking tape

Form a Hypothesis

Increasing or decreasing the number of vibrations in a second changes the pitch of a sound. For example, on a guitar, the highest notes are played when the strings vibrate the fastest. For instruments that have tubes, the length of the tube determines how quickly air inside it vibrates.

How does the length of tubes affect the pitch of sounds they make? Write your answer as a hypothesis in the form "If the tube of a wind instrument is shortened, then the pitch . . ."

_____

_____

Test Your Hypothesis

1. **Make a Model** Use scissors to cut a straw to a length of 15 cm.

2. Cut the next straw to be 1 cm shorter than the last one. Repeat this procedure until all of the straws are cut. The last straw should be 6 cm long.

Step 2

3 Lay the straws on the table. Place a piece of tape over all of the straws.

4 **Experiment** Hold the instrument to your mouth and blow across the straws to create sound.

Step 4

Draw Conclusions

5 **Explain** What do the shortest and longest pipes sound like? Was your hypothesis correct? Why or why not?

_____

_____

_____

6 **Infer** Would the 12 cm straw sound identical to the 6 cm straw if it was cut in half? Why or why not?

_____

_____

_____

Guided Inquiry

How are pitch and tension related?

Form a Hypothesis

How do you think tension in a rubber band would affect the sound it makes? Write your answer as a hypothesis in the form "If the tension in a rubber band is increased, then the pitch of the sound will . . ."

_____

_____

Test Your Hypothesis

⚠ **Be Careful.** Wear goggles. Design an experiment to investigate the effect that tension in a rubber band has on the pitch of its sound. Write out the materials you need and the steps you will follow. Record your results and observations.

_____

_____

_____

_____

Draw Conclusions

Did your results support your hypothesis? Why or why not?

_____

_____

Open Inquiry

What other variables might affect the pitch of sounds?
For example, how is sound affected by different
mediums? Determine the materials needed for your
investigation. Your experiment must be written so that
another group can complete the experiment by following
your instructions.

My hypothesis is:

_____

_____

How I can test it:

_____

_____

_____

_____

My conclusions are:

_____

_____

_____

What path does the light follow?

Materials

- masking tape
- flat mirror
- 2 pencils
- 2 erasers
- protractor

Form a Hypothesis

When you look in a mirror, you see light that travels to the mirror, bounces off, and travels to your eye. How does the angle of the light hitting the mirror compare to the angle of the light bouncing to your eye? Write your answer as a hypothesis in the form "If the angle at which light strikes a mirror decreases, then . . . "

_____

_____

Test Your Hypothesis

1. Using 2 pieces of tape, form a large letter *T*. Place the mirror upright on the top of the *T*. Stick each pencil, point down, into an eraser so that they can stand up on their own.

Step 1

2. **Experiment** Place a pencil on the left side of the *T*. Place your head on the right side. Move your head until the pencil appears to be in the center of the mirror at the top of the *T*. Now place the second pencil so that it completely blocks your view of the first pencil in the mirror.

Step 2

3. **Measure** Move the mirror and place a protractor on the top of the *T*. Find the angle between the top left of the *T* and the first pencil. This is your independent variable. Find the angle between the top right of the *T* and the second pencil. This is your dependent variable.

4. Repeat steps 2 and 3 three more times, moving the first pencil farther from the *T* each time.

Draw Conclusions

5 **Interpret Data** Look at the angles you measured. Was your hypothesis correct? Why or why not?

_____

_____

_____

Explore More

What would happen if one pencil was close to the mirror while another was far away? Would the angles change? Write a hypothesis and carry out an experiment to test it.

_____

_____

_____

© Macmillan/McGraw-Hill

Open Inquiry

Can two mirrors allow you to see around a corner?

My question is:

_____

_____

How I can test it:

_____

_____

_____

_____

My results are:

_____

_____

_____

_____

Name _____   Date _____

What happens when a ball is bounced at different angles?

Materials
- ball
- string
- tape

Form a Hypothesis

Any moving object can be reflected by a solid surface. Do the same laws that apply to the reflection of light also apply to a ball? Write a hypothesis that can be tested with the materials above. Start with "If I bounce a ball at an angle to the wall, the ball will . . ."

_____

_____

Test Your Hypothesis

1 Use a piece of string and tape to mark on the floor a path to a wall. The path should be at an angle to the wall.

2 Roll the ball along the path to the wall firmly enough to make it bounce off the wall. Use a second piece of string and tape to mark the ball's path away from the wall. Compare the two angles made by the string and the wall.

3 Repeat steps 1 and 2 several times using a different angle each time. Observe the results.

Angle 1				
Angle 2				

Draw Conclusions

4 Based on your results, what rule about the reflection of a ball from a flat surface could you write?

_____

_____

Mixing Colors

Materials

1 Divide a paper plate into six sections. Color two sections red, two sections blue, and two sections green.

- paper plate
- colored markers
- pencil
- thumbtack

2 Mount the plate on a pencil using a thumbtack.

3 **Observe** Roll the pencil between your palms to spin the wheel. What color do you see? Why?

_____

_____

_____

_____

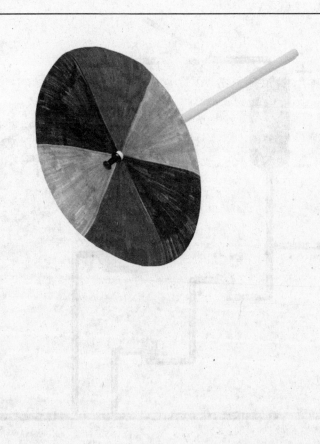

Name _____ Date _____

Which bulbs does each switch control?

Purpose

A bulb will light if there is an unbroken path through it from one end of a battery to the other. You will examine several different electrical paths with switches. You will then predict which light bulbs will be lit when a switch is opened or closed.

Procedure

1 Assemble the electric circuit as shown in the diagram, and leave all the switches open.

Materials

• three switches

• three 1.5 volt light bulbs and stands

• three 1.5 volt batteries and stands

• insulated wire with stripped leads

2 **Predict** Examine the top switch. When it closes, which bulbs will have an unbroken path from one end of a battery to another? Which bulbs will light when the switch closes? Record your prediction.

_____

_____

_____

3 **Experiment** Close the top switch and record your observations. Then open the switch.

_____

_____

4 Repeat steps 2 and 3 for the other switches.

_____

_____

Draw Conclusions

5 **Interpret Data** Look at the observations you wrote down. How many of your predictions were correct? For any that were incorrect, explain what was wrong in your thinking.

_____

_____

_____

_____

Explore More

Which switch should be closed to provide the most light
from a single bulb? What if you could close more than
one switch? Design a procedure to test which closed
switches produce the most light. Follow your procedure
and record your results.

_____

_____

_____

_____

Open Inquiry

Could a switch turn off one light and leave others on?
Think of a question about how to put a battery, wires,
three lights, and a switch together to make a circuit in
which one light can be turned off while the other two stay
lit. Make a plan and carry out an experiment to answer
your question.

My question is:

_____

_____

How I can test it:

_____

_____

_____

My results are:

_____

_____

_____

Can a lemon run my clock?

Purpose
In this activity, you will make and observe the working of a simple electric circuit.

Materials

- lemon
- penny
- galvanized nail
- 40 cm insulated copper wire, stripped at each end
- digital clock

Procedure

1 Wrap one end of a wire around the penny. Wrap one end of another wire around the nail.

2 Push the penny and the nail into one half of a lemon. Make sure the penny and the nail do not touch each other.

3 Attach the other ends of the wire to the terminals of the clock. (Hint: You may need to reverse the leads to the clock to make it run.)

Draw Conclusions

4 What path did the electricity follow to make the clock run?

_____

_____

5 What do you think would happen if you pulled the nail out of the lemon? Explain.

_____

_____

6 How long do you think the clock will run? Explain.

_____

_____

Name _____ Date _____

Measuring Electric Current

Materials
- batteries
- wire
- switch
- light bulb

1 Build a flashlight circuit using a battery, a switch, and a light bulb.

2 **Observe** Close the switch and record your findings.

_____

3 Open the circuit and add another battery. Make sure the positive end of one battery touches the negative end of the other.

4 Close the switch again. Is the light bulb the same brightness as before? Why?

_____

_____

5 **Infer** When was there more electricity flowing through the circuit? How do you know?

_____

_____

How do magnets apply forces?

Make a Prediction

Magnets push and pull on other magnets. Where on a bar magnet do you think the strongest forces are felt? Write down your prediction.

_____

_____

Test Your Prediction

1 **Observe** Lay a sealed bag containing iron filings over a bar magnet. Do the iron filings form a pattern? Draw a sketch.

Materials
• bag
• iron filings
• 2 bar magnets
• string
• meterstick
• books
• compass

2 **Experiment** Hang one magnet from a meterstick. Take another magnet and move it toward the hanging magnet. Watch how it moves. Record your observations. Repeat for each side of the bar magnet.

Farthest distance side 1 _____

Farthest distance side 2 _____

Step **2**

Name _____ Date _____

3 Place a compass at 0 cm of a
meterstick lying flat on a table.
Align the meterstick west-east.
Move a bar magnet from the
100 cm mark toward the compass.
Record at what distance the
compass first starts to move.
Repeat for each side of the bar
magnet.

Step **3**

Distance side 1 _____

Distance side 2 _____

Draw Conclusions

4 **Interpret Data** Look at all of your observations. Which
support your prediction and which disprove your
prediction? Explain. Was your prediction correct? Why or
why not?

_____

_____

_____

Explore More

Suppose you put two bar magnets on a table top in a
line, the north pole of one touching the south pole of the
other. Where do you think this double magnet would be
strongest? Design an experiment to test your prediction
and report on how accurate it was.

_____

_____

_____

Open Inquiry

What patterns would appear if magnets were placed in other positions? Think of a question about how to put magnets together to make different patterns. Make a plan and carry out an experiment to answer your question.

My question is:

_____

_____

How I can test it:

_____

_____

_____

My results are:

_____

_____

_____

How does a material become a magnet?

Materials

- bar magnet
- sealed plastic bag with iron filings

Purpose

The atoms in magnetic materials such as iron are tiny magnets themselves. In nature, a magnetic material becomes a magnet (is magnetized) as it hardens from molten magma into rock. Earth acts like a giant magnet that causes the atom-sized magnets to point in the same direction. The greater the number of these tiny magnets that point in the same direction, the stronger the magnet. You can model this process with iron filings.

Procedure

1 Pass one pole of a bar magnet slowly over a small plastic bag with iron filings. What did you observe?

_____

2 Pass the magnet over the iron filings several more times. What happened?

_____

3 Shake the bag of iron filings, and repeat steps 1 and 2.

Draw Conclusions

4 How were the iron filings affected by the magnet?

_____

5 What happened when you shook the bag of filings?

_____

6 How could the iron filings become part of a permanent magnet?

_____

_____

Building an Electromagnet

Materials

- 2 pieces of insulated copper wire 1 m and 2 m in length
- pencil
- compass
- battery
- small steel paper clips
- nail

1 Coil a length of insulated wire around a pencil 25 times. Remove the pencil.

2 **Observe** Place a compass right under the wire coil. Turn the coil so that it is crosswise to the compass needle. Touch the ends of the wire to a battery. Write down what you observe.

_____

3 Hold the ends of the wire to the battery and try to pick up small steel paper clips with the coil. What's the largest chain of paper clips you can lift?

Number of clips _____

4 Repeat steps 2 and 3 with a nail inserted into the coil. Then repeat the test with a longer coil.

Number of clips with nail _____

Number of clips with longer

coil _____

_____

_____

5 **Interpret Data** How would you make the strongest electromagnet with the materials you used?

_____

_____

_____

_____

Name _____ Date _____

How are electric current and electromagnets related?

Materials

- wire with leads
- 3-inch nail
- masking tape
- battery
- battery holder
- compass

Form a Hypothesis

A magnetic field is produced when a current is flowing in a circuit. An electromagnet can be produced in this way. Electromagnets produce a magnetic field similar to bar magnets. When the current stops, the magnetic field disappears.

Each electromagnet has a north and a south pole. A compass needle also has a north and south pole. The compass needle will point to the appropriate poles of other magnets. How do you think the direction of electric current affects the poles of the electromagnet? Write a hypothesis in the form "If the direction of electric current is reversed, then the poles on an electromagnet . . ."

_____

_____

Test Your Hypothesis

1. Coil the wire around the nail 35 times clockwise toward the flat end. Leave about 10 cm of straight wire at both ends.

2. Find the straight part of the wire near the flat side of the nail. Connect that end of the wire to the positive side of the battery.

Step 1

3 Lay the compass on the flat end of the nail. Press the unconnected wire to the negative side of the battery. Record what happens.

Step 3

_____

_____

4 Find the wire from the flat end of the nail. Disconnect it from the positive side of the battery and connect it to the negative side. Keep the compass on the flat end of the nail. Press the other end of the wire to the positive side of the battery. Record what happens.

_____

_____

Draw Conclusions

5 **Infer** Where did the compass point in step 3 and step 4? What do you think happened to the poles of the electromagnet?

_____

_____

_____

6 **Communicate** Draw a picture of the electromagnet before and after the current was reversed. Mark to which side of the battery the wires were connected. Label the poles of the electromagnet as north or south.

Electromagnet Before Current Was Reversed

Electromagnet After Current Was Reversed

Guided Inquiry

How is an electromagnet affected by the direction of its coils?

Form a Hypothesis

Are the poles of an electromagnet only dependent on electric current? How does the direction in which a coil is wound affect an electromagnet? Write your answer as a hypothesis in the form "If the direction in which a coil is wound is reversed, then the poles of the electromagnet . . ."

_____

_____

Test Your Hypothesis

Design an experiment to investigate the effect changing the direction of coils will have on an electromagnet. Write out the materials you need and the steps you will follow. Record your observations.

_____

_____

_____

_____

© Macmillan/McGraw-Hill

Draw Conclusions

Did your results support your hypothesis? Explain.

_____

_____

_____

_____

Open Inquiry

What can you learn about electromagnets? For example, how are electromagnets used in electric motors? Determine the materials needed for your investigation. Your experiment should be written so that another group can complete it by following your instructions.

My question is:

_____

_____

How I can test it:

_____

_____

_____

_____

_____

My conclusions are:

_____

_____

_____

© Macmillan/McGraw-Hill

Breathe Deeply

The respiratory system is an organ system that takes in oxygen and removes carbon dioxide. The respiratory system is made up of the airway, lungs, and diaphragm. The diaphragm is a muscle that controls the physical motion of breathing. When you inhale, the diaphragm contracts to make room for the lungs to expand and fill with air. When you exhale, the diaphragm expands and pushes air out of the lungs.

Materials

- 2-liter plastic bottle
- balloon
- plastic wrap
- tape
- construction paper

Purpose

Your task is to create a model of a lung and the diaphragm to show how they work.

Form a Hypothesis

What would happen if there was a hole in the diaphragm? State your hypothesis in the form of an "if", "then" statement. *("If the diaphragm has a hole in it, then . . .")*

_____

_____

_____

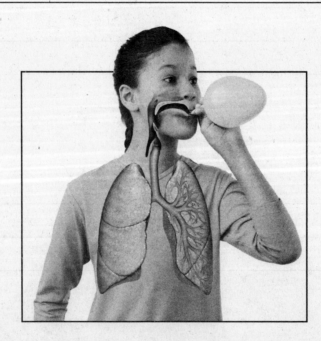

Test Your Hypothesis

1 ⚠ **Be Careful.** Cut off the top half of the 2-liter bottle and attach the balloon to the neck. Then push the balloon into the bottle top.

2 Stretch plastic wrap over the bottom of the bottle and tape it to the bottle tightly, making an airtight seal.

3 Fold a strip of construction paper in half and securely tape it to the bottom of the plastic wrap so that a handle is created.

4 Observe the model as you pull and push on the handle gently.

5 Use a pencil to poke a small hole in the plastic wrap on the bottom of the bottle. Then pull and push on the handle of the bottle again.

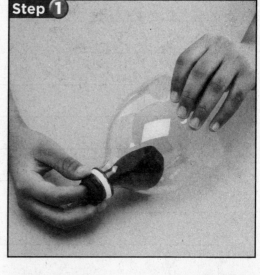

Step 1

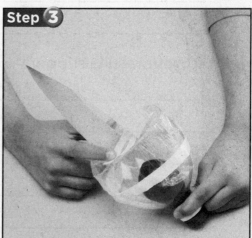

Step 3

Draw Conclusions

6 What did you observe?

_____

_____

_____

_____

_____

_____

_____

_____

7 Based on your results, what is your conclusion?

_____

_____

_____

_____

_____

_____

_____

_____

8 Did your results support your hypothesis?

_____

_____

_____

_____

_____

Critical Thinking

1 Do plants also take in oxygen and give off carbon dioxide?

_____

_____

_____

_____

_____

_____

_____

_____

_____

2 Where does oxygen go once it enters the lungs?

_____

_____

_____

_____

_____

_____

_____

_____

_____

_____

© Macmillan/McGraw-Hill

Name _____ Date _____

Salty Cells

The cell is the smallest unit of a living thing and it is made up mostly of water. In fact, 2/3 of the body's water is found in its cells. There are many different kinds of cells in your body such as muscle, skin, brain, blood, liver, and stomach cells. All cells are surrounded by a cell membrane. The cell membrane controls what moves in and out of the cell. Osmosis is a very important process in regulating the amount of water in the cells. Osmosis is the movement of water through the cell membrane from an area of low concentration of a substance to an area of high concentration of a substance. Water molecules will move in or out of cells depending on where there is a higher concentration, or amount, of a substance such as salt.

Materials
• potato slices
• 2 bowls or petri dishes
• salt
• water
• tablespoon

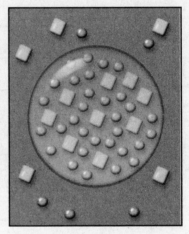

Purpose
Your task is to test the concentration of potato cells in salt water.

Form a Hypothesis
What would happen if you put a slice of potato (which contains lots of cells) in a dish with very salty water? State your hypothesis in the form of an "if", "then" statement. *("If a slice of potato is put in a dish with salty water, then the potato will . . .")*

_____

_____

_____

_____

Test Your Hypothesis

1 Fill two petri dishes with enough water to cover several potato slices.

2 Add two tablespoons of salt to one of the dishes.

3 Put 3 potato slices in the dish with the salt and 3 slices in the dish without the salt.

4 Let the potato slices soak for at least 20 minutes.

5 Drain the remaining water from the petri dishes. Observe the potato slices.

Step 2

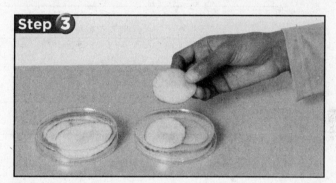

Step 3

Draw Conclusions

6 What did you observe?

_____

_____

_____

_____

_____

_____

_____

_____

Everyday Science

Name ____________________ Date __________

7 Based on your results, what is your conclusion?

Critical Thinking

1 Why do potato chips make you thirsty?

_____

_____

_____

_____

_____

_____

_____

_____

2 Why did you use two dishes with water and potato slices?

_____

_____

_____

_____

_____

_____

_____

_____

Leaf Pigments

Leaves are usually green because they contain chlorophyll, a chemical in plant cells with a green pigment, or color. This chemical is very important. It absorbs sunlight so that the plant can make its own food. In the fall there is less sunlight, so plants make less chlorophyll. As result, the other pigments in the leaves become more visible. This is what gives leaves their fall colors.

Scientists use chromatography to separate different kinds of substances by their pigments. A fluid travels up special chromatography paper and carries small substances with it. The liquid, such as alcohol, travels up the paper at different speeds depending on the weight of the substances. In this experiment, bands of color will appear on the chromatography paper to show which pigments are present.

Materials
• leaves
• plastic cups
• rubbing alcohol
• popsicle stick
• 2 beakers with warm water
• chromatography paper

Purpose
Your task is to use chromatography to test leaves.

Form a Hypothesis
Which pigments are present in leaves? State your hypothesis in the form of an "if", "then" statement. *("If _____ colored leaves are mixed with alcohol, then _____ will be seen on the chromatography paper.")*

_____
_____
_____
_____
_____
_____
_____

Test Your Hypothesis

1. Tear 2 to 4 green leaves into small pieces and add them to the small plastic cup. Tear 2 to 4 yellow or red leaves and add them to the other small plastic cup.

2. Add rubbing alcohol to both cups so that the leaves are covered.

3. Mix the leaves and the alcohol solution with the popsicle stick for 2 minutes.

4. Place each cup in a beaker half full with warm tap water. Leave the cups in the beakers until the rubbing alcohol in the cups changes color. If the water cools off, replace it with more warm water.

5. Carefully remove the cups from the beakers.

6. Put a strip of chromatography paper in each cup and leave it for 1 hour.

Step 4

Step 6

Draw Conclusions

7. What did you observe?

_____

_____

_____

_____

_____

_____

_____

8 Based on your results, what is your conclusion?

_____

_____

_____

_____

_____

_____

_____

_____

_____

_____

_____

Critical Thinking

1 Why do many trees lose their leaves in the winter?

_____

_____

_____

_____

_____

_____

_____

_____

_____

_____

2 When do evergreen trees lose their leaves?

_____

_____

_____

_____

_____

_____

_____

_____

_____

_____

Name _____ Date _____

Stomach Acid

The digestive system is important for taking in, breaking down, and absorbing food. Food travels from your mouth through the esophagus to your stomach. Your stomach has an enzyme called pepsin. Pepsin is important because it starts breaking down protein molecules into smaller pieces. This is necessary for their absorption by the small intestine. However, pepsin is only active when it is in the presence of an acid called hydrochloric acid (HCl). This acid is secreted by special cells in the stomach wall.

Materials

- shallow dish
- gelatin
- drinking straw
- toothpicks
- pepsin
- plastic cup
- eye-dropper
- water

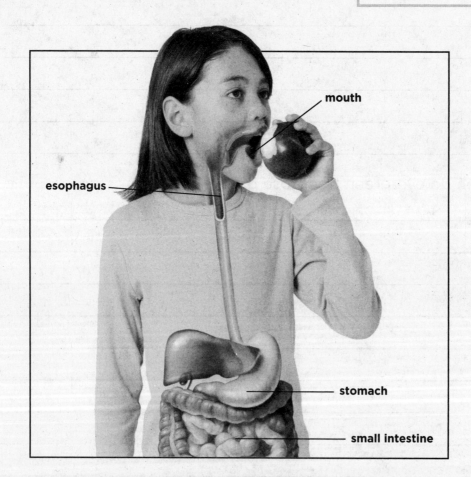

mouth

esophagus

stomach

small intestine

Purpose

You task is to digest a protein (gelatin) using pepsin.

Form a Hypothesis

How does pepsin break down proteins? State your hypothesis in the form of an "if", "then" statement. ("If the pepsin is added to the gelatin, then the gelatin will . . .")

_____

_____

_____

_____

Test Your Hypothesis

1 Prepare the gelatin and put it in the shallow dish.

2 Put the dish in the refrigerator until the gelatin sets.

3 Use the straw to poke 10 holes in the gelatin (spread the holes out).

Step 3

4 Use the toothpick to take the gelatin plugs out of the holes.

5 Break open the pepsin capsule and put the contents into the cup. Add 1 teaspoon of water to the cup and mix. Draw up the pepsin mixture into the eye-dropper.

6 Fill up 5 of the holes with the pepsin mixture. Keep track of which holes have the pepsin. Let it sit for a couple of hours or overnight.

Step 6

© Macmillan/McGraw-Hill

Draw Conclusions

7 What did you observe? What happened?

_____

_____

_____

8 Based on your results, what is your conclusion?

_____

_____

_____

_____

_____

_____

_____

_____

_____

_____

_____

_____

_____

Critical Thinking

1 Why didn't you fill up all 10 holes with the pepsin mixture?

_____

_____

_____

_____

_____

_____

_____

_____

2 What else was in the pepsin capsule besides the pepsin enzyme?

_____

_____

_____

_____

_____

_____

_____

_____

Name _____   Date _____

Growing Plants Without Watering Them

Water exists in three phases: liquid, solid, and gas. Water can change from one phase to another. Condensation, evaporation, and precipitation are processes of the water cycle. How can you set up an environment in which the water cycle provides water for plants so you do not need to add any extra water to the environment?

Materials

- glass jar
- charcoal
- soil
- seed mixture
- spoon
- water

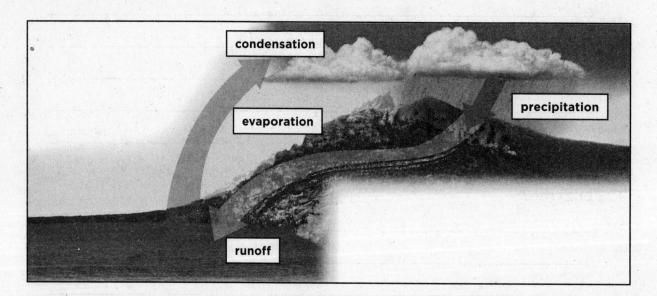

Purpose

Make a model of the water cycle that will provide water for plants.

Procedure

1. Place the glass jar sideways on a table. Using the spoon, put a layer of charcoal on the bottom side of the jar.

2. Pour water onto the soil until the soil is soaking wet. Using the spoon, place the wet soil on top of the charcoal.

© Macmillan/McGraw-Hill

3 Place ten seeds on top of the soil.

4 Cover the seeds with a thin layer of soil.

5 Using a spoon, add more water to the soil.

6 Screw the lid onto the jar.

7 **Observe** After two days, where do you see water in the jar? What happens to the seeds?

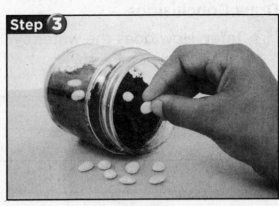

Step **3**

_____

_____

_____

_____

_____

_____

_____

_____

_____

_____

© Macmillan/McGraw-Hill

Draw Conclusions

8 **Infer** How does the water cycle in the jar resemble the water cycle on Earth?

_____

_____

_____

_____

_____

9 **Communicate** Did all ten of your plants grow? Write a report about your investigation.

_____

_____

_____

_____

_____

_____

_____

_____

_____

_____

_____

_____

_____

_____

_____

_____

Critical Thinking

1 Would you expect to see fog or clouds form in your jar?

_____

_____

_____

_____

_____

_____

_____

_____

_____

2 Do you think the amount of water on Earth has changed?

_____

_____

_____

_____

_____

_____

_____

_____

_____

_____

Name _____ Date _____

The Humidity of the Air

You learned that air contains some amount of water vapor. The amount of water vapor in air is called humidity. How can you measure humidity?

Purpose
You will make an instrument to measure humidity.

Procedure

1 Pull a cotton ball over the bulb of one thermometer. Tape that thermometer to one side of the carton. Tape the second thermometer to another side of the carton.

2 Using a hole punch, punch two holes in the top of the carton.

3 Thread a long piece of string through the holes. Tie the ends of the string together to form a large loop.

4 While holding the carton, pour water on the cotton ball.

5 ⚠ **Be Careful.** While holding onto the string, swing the carton gently around you for one minute.

6 **Record Data** Write down the temperatures on the two thermometers. Which thermometer has a lower temperature?

thermometer 1 _____

thermometer 2 _____

Materials

- pint-sized carton
- water
- two thermometers
- cotton ball
- string
- clear tape
- hole punch
- container

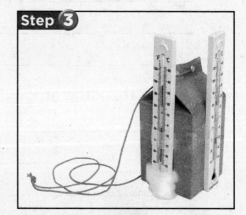
Step 3

Step 5

Draw Conclusions

7 **Infer** Why does one thermometer have a lower temperature?

_____

_____

_____

_____

_____

_____

_____

_____

8 **Use Numbers** Use the chart below to calculate the humidity.

Relative Humidity

Dry Bulb Temperature Minus Wet Bulb Temperature

	1	2	3	4	5	6	7	8	9	10	15	20	25
30	89	78	67	56	46	36	26	16	6				
35	91	81	72	63	54	45	36	27	19	10			
40	92	83	75	68	60	52	45	37	29	22			
45	93	86	78	71	64	57	51	44	38	31			
50	93	87	80	74	67	61	55	49	43	38	10		
55	94	88	82	76	70	65	59	54	49	43	19		
60	94	89	83	78	73	68	63	58	53	48	26	5	
65	95	90	85	80	75	70	66	61	56	52	31	12	
70	95	90	86	81	77	72	68	64	59	55	36	19	3
75	96	91	86	82	78	74	70	66	62	58	40	24	9
80	96	91	87	83	79	75	72	68	64	61	44	29	15
85	96	92	88	84	80	76	73	70	66	62	46	32	20
90	96	92	89	85	81	78	74	71	68	65	49	36	24
95	96	93	89	86	82	79	76	72	69	66	52	38	28

Dry Bulb temperature (degrees F)

Humidity _____

9 **Communicate** Write a report about your investigation. Include your results. What humidity measurement did you get?

_____

_____

_____

_____

_____

_____

_____

_____

_____

_____

_____

_____

_____

_____

_____

_____

_____

_____

_____

Critical Thinking

1 How does having one wet and one dry thermometer help you to figure out the amount of water in the air?

_____

_____

_____

_____

_____

_____

_____

_____

2 If you measured the humidity on a sunny day and a rainy day, on which day would you expect the humidity to be higher? Why?

_____

_____

_____

_____

_____

_____

_____

_____

Name _____ Date _____

Gravity and Inertia

An orbit is the path of an object, such as a planet, through space around another object. There are two forces that keep a planet in an orbit. One force is gravity, which pulls the planet toward the Sun. The other force is inertia, a force created by the planet's speed. When these forces are balanced the planet moves in an orbit. How can you use these forces to keep a marble inside an upside-down soda bottle?

Materials
• 2-liter plastic bottle
• marble

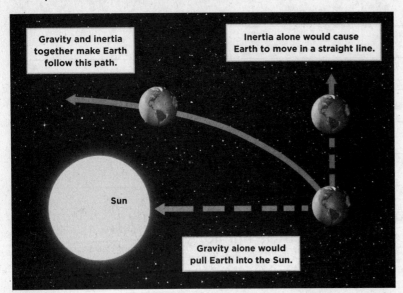

Gravity and inertia together make Earth follow this path.

Inertia alone would cause Earth to move in a straight line.

Sun

Gravity alone would pull Earth into the Sun.

Purpose

This activity will demonstrate that inertia will keep an object in orbit.

Procedure

1. Drop the marble into an empty plastic bottle. Holding the bottle by the top, swirl the bottle so the marble starts rolling in a circle around the bottom.

2. ⚠ **Be Careful.** While swirling the bottle rapidly and holding the bottle by the top, quickly flip the bottle upside-down, and continue to swirl the bottle.

Step 1

© Macmillan/McGraw-Hill

Use with **Activity Flipchart p. 81**

3 **Observe** What path is the marble taking?

_____

_____

_____

_____

_____

_____

4 Decrease the speed with which you are swirling the
bottle. What happens?

_____

_____

_____

_____

_____

_____

5 Have another student hold his or her hands under the
opening of the bottle. Then stop swirling the bottle
completely. What happens?

_____

_____

_____

_____

_____

_____

Name _____ Date _____

Draw Conclusions

6 What does the marble represent in this model?

_____

_____

_____

_____

7 How does this model resemble what would happen between a planet and a Sun if the speed of the planet slowed?

_____

_____

_____

_____

8 Write a report about your investigation. Include your results. Where could you see the effects of inertia and gravity in this experiment?

_____

_____

_____

_____

_____

_____

_____

Critical Thinking

1 What would happen to a planet if there were no inertia?

_____

_____

_____

_____

_____

_____

_____

_____

_____

2 How fast must a satellite travel to stay in orbit?

_____

_____

_____

_____

_____

_____

_____

_____

_____

Name _____   Date _____

Electroplating

Electroplating is the process by which a thin layer of metal is attracted by electric charges to the surface of another metal and deposited there. For example, pennies are actually copper-plated zinc, zinc with a thin layer of copper covering it. Over time, pennies become dull because the copper is reacting with the air to form copper oxide. Copper oxide is dull and greenish. Pennies can be cleaned by putting them in an acid such as vinegar and salt. This solution will remove the copper oxide. The copper in the solution can now react with another metal.

Materials

- vinegar
- salt
- cup
- galvanized nails
- 10 pennies

Purpose

Your task is to see if you can plate another metal with copper.

Form a Hypothesis

What do you think would happen if you put 10 pennies and 2 galvanized nails in a cup of vinegar and salt? State your hypothesis in the form of an "if", "then" statement. *("If I put 10 pennies and 2 galvanized nails in a vinegar and salt solution, then . . .")*

_____

_____

_____

_____

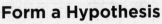

Name _____ Date _____

Test Your Hypothesis

1 Mix 60 mL of vinegar and 1 teaspoon of salt in a cup.

2 Put 10 dull pennies in the cup.

3 After 10–15 minutes, take 5 of the pennies out when shiny.

4 **Observe** Add two nails to the cup and wait 10–15 minutes.

Step **1**

Step **2**

Draw Conclusions

5 What did you observe?

_____

_____

_____

_____

_____

_____

_____

_____

_____

6 Based on your results, what is your conclusion?

_____

_____

_____

_____

_____

_____

_____

_____

_____

_____

7 Write a report about your investigation.

_____

_____

_____

_____

_____

_____

_____

_____

_____

_____

_____

_____

© Macmillan/McGraw-Hill

Critical Thinking

1 Can you think of a better way to copper plate an object?

_____

_____

_____

_____

_____

_____

_____

_____

_____

2 Can you think of additional examples of electroplating?

_____

_____

_____

_____

_____

_____

_____

_____

_____

Structured Inquiry

How does temperature affect the rate of water transport through plant stems?

Water Race

Form a Hypothesis

Vascular plants transport materials through specialized tissues. Water is transported from the roots up to the leaves through xylem tissue in the stem. How does temperature affect the rate at which the water is transported through the xylem of a plant stem? Write your answer as a hypothesis in the form *"If the water temperature is decreased, then . . ."*

Materials
• 6 celery stalks with leaves
• water
• ice
• blue food coloring
• 2 500 mL beakers
• 2 thermometers
• ruler
• paper towels
• scissors

_____

_____

_____

_____

_____

_____

Test Your Hypothesis

1 Fill one beaker, labeled Beaker A, with room temperature water (about 25 degrees Celsius) and the other beaker, labeled Beaker B, with ice water (about 10 degrees Celsius).

2 Add 10 drops of blue food coloring to each beaker.

3 Place a thermometer in each beaker. Be sure to add ice to Beaker B if the temperature goes above 15 degrees celsius.

4 Place 3 stalks of celery in each beaker.

Step **4**

Experiment

5 **Observe** After 15 minutes remove one stalk of celery from Beaker A and one from Beaker B and place them on a paper towel. Use the scissors to carefully scrape the celery stalks to expose the xylem tissue. What do you observe?

_____

_____

_____

_____

6 **Measure** Use the ruler to measure how far the water has traveled up the celery stalk. Record your measurements.

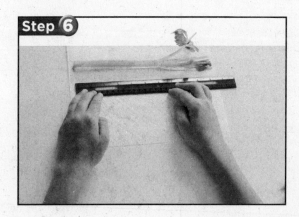

Step **6**

7 Repeat steps 5 and 6 every 15 minutes for 30 more minutes.

Measurement 1: Celery Stalk A (ice water) _____

Celery Stalk B (warm water) _____

Measurement 2: Celery Stalk C (ice water) _____

Celery Stalk D (warm water) _____

Measurement 3: Celery Stalk E (ice water) _____

Celery Stalk F (warm water) _____

8 **Record Data** Use your data to make a bar graph. Put the celery data along the bottom or horizontal side of the graph. Put the water height measurements on the left, or vertical side of the graph. Draw a bar for each celery stalk.

Communicate Your Results

Have a class discussion and share your results and graphs.
What did you find out? Use your data to answer the
questions.

▶ What was the controlled variable in this experiment?
the dependent variable? the independent variable?

_____

_____

_____

_____

_____

▶ **Interpret Data** Did you observe a correlation between
the rate of water transport and the temperature of the
water?

_____

_____

_____

_____

_____

▶ Did your results support your hypothesis?

_____

_____

_____

_____

_____

Guided Inquiry

Salty Celery

Form a Hypothesis

You have already tested the effects of water temperature in plant stems. What other variables will slow or increase the rate that water flows through a plant stem? Will dissolving substances, such as salt or sugar, change the water flow rate? Write your answer as a hypothesis in the form *"If salt is dissolved in water, then . . ."*

Materials
• 6 celery stalks with leaves
• 2 500 mL beakers
• salt
• tablespoon
• blue food coloring
• scissors
• ruler
• paper towels

_____

_____

_____

_____

_____

_____

_____

_____

_____

_____

_____

_____

_____

© Macmillan/McGraw-Hill

Test Your Hypothesis

1 Fill 2 beakers with water and add 10 drops of blue food coloring to each beaker.

2 Pour 2 tablespoons of salt into one of the beakers.

3 Place 3 celery stalks in each beaker.

Step **3**

4 **Observe** After 15 minutes remove a stalk of celery from each beaker and place them on a paper towel.

5 Use the scissors to carefully scrape the celery stalks to expose xylem tissue.

6 **Measure** Use the ruler to measure how far the water has traveled up the celery stalks. Record your measurements.

7 Repeat steps 4 through 6 for the remaining celery stalks.

Measurement 1: Celery Stalk A (fresh water) _____

Celery Stalk B (salt water) _____

Measurement 2: Celery Stalk C (fresh water) _____

Celery Stalk D (salt water) _____

Measurement 3: Celery Stalk E (fresh water) _____

Celery Stalk F (salt water) _____

Communicate Your Results

Work in groups of 4 to 8 and discuss what you found out about water transport in plants.

▶ How did adding salt affect the transportation rate of water in the celery?

_____

_____

_____

_____

▶ How did your results compare with the group?

_____

_____

_____

_____

_____

_____

▶ Did your results support your hypothesis? Why or why not?

_____

_____

_____

_____

_____

_____

_____

© Macmillan/McGraw-Hill

Open Inquiry

Water Transport Systems in Other Plants

Invent and test other ways to explore the xylem tissue in plants. Design and perform an experiment. Ask a question, make a prediction, record your data, and communicate your findings. Make a poster to show what you did and what you found out. Here are some ideas to get you started.

▶ What is the structure of xylem tissue like in other plants? Does this tissue look like and function in the same way as xylem in celery stems? You could test the roots and stems of other vegetables like carrots or potatoes or flowers like carnations or daisies.

_____

_____

_____

_____

_____

_____

_____

© Macmillan/McGraw-Hill

Name _____ Date _____

What effect does temperature have on the formation of clouds?

Structured Inquiry

Temperature and Cloud Formation

Ask Questions
Clouds form when water vapor condenses. What effect does temperature have on the formation of clouds?

Form a Hypothesis
If you have two sources of water vapor, such as two lakes, and you lower the temperature of the air above one of the sources more than you lower the temperature of the air above the other one, over which source would you expect a more visible cloud to form? Write your answer as a hypothesis in the form *"If the temperature of the air above one source of water vapor is lowered more than the other, then . . ."*

© Macmillan/McGraw-Hill

Materials
• hot water
• 4 identical clear containers
• 2 sealable storage bags
• 6 ice cubes
• clear tape

_____

_____

_____

_____

_____

_____

_____

_____

_____

Use with **Activity Flipchart pp. 85-86**

Test Your Hypothesis

Step 3

1. Label 2 of the containers "Bottom." Label the other two containers "1" and "2."

2. Put container 1 in a cool place for about ten minutes.

3. Place three ice cubes in each sealable storage bag and seal the bags.

4. Fill the two containers labeled "Bottom" with equal amounts of hot water.

Step 7

5. **Make a Model** Place container 1 upside down on top of one of the Bottom containers. Tape the two containers together.

6. **Make a Model** Place container 2 upside down on top of the other Bottom container. Tape the two containers together.

7. Put the storage bags full of ice cubes on top of the containers.

8. **Observe** What did you observe?

_____

_____

_____

_____

_____

_____

_____

_____

Draw Conclusions

▶ **Record Data** Draw what you observed in both setups.

[two blank boxes for drawing]

▶ What is the independent variable in this experiment?

_____

_____

_____

▶ How was the independent variable changed to collect information?

_____

_____

_____

▶ Do clouds appear more visible when they form in colder air?

_____

_____

_____

© Macmillan/McGraw-Hill

Guided Inquiry

Presence of Dust in the Air and Cloud Formation

Ask Questions

Clouds form because as the air cools, the water vapor contained in it begins to condense. Is temperature the only factor that controls how visible a cloud is when it forms?

Form a Hypothesis

What effect do you think the presence of dust in the air has on cloud formation? Write your answer as a hypothesis in the form *"If there is dust in the air, then . . ."*

_____

_____

_____

_____

_____

_____

_____

_____

_____

_____

_____

Materials
• 2 large glass jars
• 2 large balloons
• scissors
• chalk dust
• large rubber bands
• markers

Name _____   Date _____

Test Your Hypothesis

1 Cut the open ends off of the two balloons.

2 Spray a few drops of warm water on the insides of each jar.

3 Stretch one balloon over the top of the first jar. Wrap rubber bands around the balloon and the top of the jar so the balloon forms a tight seal.

Step **3**

4 Clap a dusty chalkboard eraser over the top of the second jar, so some dust falls to the bottom of the jar. Then seal the jar with the second balloon and more rubber bands. Label this jar "Chalk."

5 Place both jars in a cool place for about 10 minutes.

6 **Observe** Remove the jars and compare them. Then pull on each of the two balloons. What do you see?

_____

_____

_____

_____

_____

_____

_____

_____

_____

_____

© Macmillan/McGraw-Hill

Draw Conclusions

1 Where did you see clouds form? Which was the most visible cloud?

_____

_____

_____

_____

_____

_____

_____

_____

2 **Predict** What would happen if you repeated this experiment with cold water in the container?

_____

_____

_____

_____

_____

_____

_____

_____

_____

Name _____ Date _____

Open Inquiry

Clouds Can Be Found Almost Anywhere

What other questions do you have about cloud formation? What other ways can clouds form? Come up with a testable question and perform an experiment to answer the question. Here are some ideas to get you started.

▶ What happens if you put a 2-liter bottle filled with water and smoke in a refrigerator?

▶ Can you make a cloud in your bathroom at home?

▶ Look at a weather map and find places where warm and cold air meet. Predict how often you would observe clouds in those places. Then compare your prediction with the cloudiness that was observed.

_____

_____

_____

_____

_____

© Macmillan/McGraw-Hill

Use this page for any notes you made about the experiments.

_____

_____

_____

_____

_____

_____

_____

_____

_____

_____

_____

_____

_____

_____

_____

_____

_____

_____

_____

Name _____ Date _____

Structured Inquiry

What changes take place during a fizzing reaction?

Exploring a Fizzing Reaction

Ask Questions

Only chemical changes form new materials. Physical changes do not. You can look for changes in composition or color when deciding if a new material has formed. What happens when an antacid fizzing tablet is placed in tap water? Are there any changes? Are they chemical or physical? Are new materials formed or not?

Make a Prediction

What do you think will happen when you put an antacid tablet into water with Phenol Red?

_____

_____

_____

_____

_____

_____

_____

_____

_____

_____

_____

Materials
• antacid tablet
• clear plastic cup
• pH indicator (Phenol Red) with eyedropper
• graduated cylinder or measuring cup
• plastic spoon

Test Your Prediction

1 Put 25 mL of cold tap water into a plastic cup. Add 5 drops of phenol red (or another pH indicator) to the water. Add enough phenol red so that there is a noticeable color to the water.

2 Drop an antacid tablet into the water. Do you observe any changes of color or other changes? What can you infer about the formation of new materials? What kind of changes took place?

Step **2**

_____

_____

_____

_____

_____

_____

_____

_____

_____

_____

_____

_____

_____

_____

Name _____ Date _____

Draw Conclusions

▶ How did the fizzing antacid tablet change?

_____

_____

▶ How did the pH change when the tablet was added?

_____

_____

▶ How would the pH change if you stirred the solution with a spoon 10 minutes after the reaction?

_____

_____

_____

▶ After the reaction ends, pour the liquid into a plate and allow the water to evaporate. Is the residue the same as the original tablet? How would you know? Where did the tablet go?

_____

_____

_____

_____

_____

▶ **Infer** Was this a chemical or physical reaction? What are the reasons for your inference?

_____

_____

_____

_____

_____

Communicate Your Results

Have a class discussion and share your results with the other students. Use your data to answer the questions.

▶ What type of reaction took place?

_____

_____

_____

_____

_____

▶ How can you support your conclusions?

_____

_____

_____

_____

_____

_____

▶ What were your classmates' conclusions? How are they the same or different from yours?

_____

_____

_____

_____

_____

Guided Inquiry

How Long Can It Fizz?

Ask Questions

Would crushing the tablet into powder make the amount of time the reaction takes place increase, decrease, or remain the same? What other factors might change the amount of time the reaction takes?

Make a Prediction

Predict what will happen to the length of the fizzing reaction when you put a crushed fizzing antacid tablet into water with phenol red.

_____

_____

_____

_____

_____

_____

_____

_____

_____

_____

_____

_____

_____

Materials
• 1 antacid tablet
• clear plastic cup
• pH indicator (Phenol Red) with eyedropper
• graduated cylinder or measuring cup
• stopwatch
• thermometer

Test Your Prediction

1 Work with a partner. Break an antacid tablet in half. Wrap one half-tablet in a sheet of paper and crush it into powder. Leave the remaining half uncrushed.

2 Add 15 mL of cold tap water and 2 drops of pH indicator to each cup. Start timing as you drop the crushed tablet into one cup and the uncrushed tablet into the other cup. The cup with the uncrushed tablet is the experimental control.

3 How does crushing affect the fizzing time? Identify the controlled and dependent variables.

_____

_____

_____

_____

4 What factor would you like to check next? Does water temperature affect fizz time? Does the pH indicator affect the fizz?

Step **4**

5 Devise an experiment to test your question. Identify the controlled and dependent variables.

_____

_____

_____

_____

_____

Communicate Your Results

Work in groups of 4 to 8 and discuss what you found out about fizzing times.

▶ Write a report of one of your investigations. Discuss how your results compared to your classmates.

_____

_____

_____

_____

_____

_____

_____

_____

_____

_____

_____

_____

_____

_____

_____

_____

_____

_____

_____

_____

Open Inquiry

More Fun with Fizzing

Design and perform an experiment to answer your question. Make a prediction, perform an experiment to test it, record your data, and communicate your findings. What are the independent and dependent variables, and what is your control? Write a report of your experiment. Include step-by-step directions so others can repeat what you did. Here are some ideas to get you started:

▶ How much fizz will we get if we use only 6 drops of water?

▶ Does the temperature of the liquid in the cup change when the mixture fizzes?

▶ Is the gas given off like bubbles in soda pop? How could we test our idea?

_____

_____

_____

_____

_____

_____

_____

_____

_____

_____

_____

_____